Biochemical Basis of Plant Breeding

Volume I
Carbon Metabolism

Editor

Carlos A. Neyra, Ph.D.
Associate Professor of Plant Physiology
Biochemistry and Microbiology Department
Rutgers University
New Brunswick, New Jersey

CRC Press
Taylor & Francis Group
Boca Raton London New York

CRC Press is an imprint of the
Taylor & Francis Group, an **informa** business

CRC Press
Taylor & Francis Group
6000 Broken Sound Parkway NW, Suite 300
Boca Raton, FL 33487-2742

Reissued 2019 by CRC Press

© 1985 by Taylor & Francis Group, LLC
CRC Press is an imprint of Taylor & Francis Group, an Informa business

No claim to original U.S. Government works

A Library of Congress record exists under LC control number:

Publisher's Note
The publisher has gone to great lengths to ensure the quality of this reprint but points out that some imperfections in the original copies may be apparent.

Disclaimer
The publisher has made every effort to trace copyright holders and welcomes correspondence from those they have been unable to contact.

ISBN 13: 978-0-367-26094-1 (hbk)
ISBN 13: 978-0-367-26095-8 (pbk)
ISBN 13: 978-0-429-29143-2 (ebk)

Visit the Taylor & Francis Web site at http://www.taylorandfrancis.com and the
CRC Press Web site at http://www.crcpress.com

THE EDITOR

Carlos A. Neyra, Ph.D., is an Associate Professor of Plant Physiology in the George H. Cook College of Agriculture and Environmental Sciences at Rutgers University, New Brunswick, New Jersey. Dr. Neyra received his B.S. degree in Agronomy in 1965 from the Universidad Nacional de Tucuman in Argentina and his Ph.D. degree in Plant Physiology in 1974 at the University of Illinois at Urbana-Champaign.

In addition to being a prolific writer in the fields of Agronomy, Agricultural Biochemistry, and Plant Physiology, Dr. Neyra is widely known internationally for his activities as a Teacher, Researcher, and Scientific Advisor in several countries including Argentina, Brasil, Dominican Republic, Panama, Peru, and the United States.

Dr. Neyra is a tenured member of the Rutgers faculty with full-member status in the following graduate programs: Botany and Plant Physiology; Horticulture; Microbiology; and Soils and Crops. He teaches formally both undergraduate and graduate courses of Plant Physiology and has served in numerous University Committees. He is a reviewer for a number of referred journals including *Plant Physiology, Canadian Journal of Botany, Canadian Journal of Microbiology, Soil Sciences,* and others. He has also collaborated as a peer reviewer for various granting agencies (USDA, NAS, NSF, AID, BARC, BOSTID, etc.). Dr. Neyra was elected full-member of the honorary societies of Sigma Xi and Gamma Sigma Delta in 1974.

CONTRIBUTORS

James T. Bahr, Ph.D.
Research Manager
Rhone-Poulenc, Inc.
Agrochemical Division
Monmouth Junction, New Jersey

Clanton C. Black, Jr., Ph.D.
Research Professor of Biochemistry
Biochemistry Department
Boyd Graduate Studies Research Center
University of Georgia
Athens, Georgia

Charles D. Boyer, Ph.D.
Associate Professor of Plant Breeding
  and Genetics
Department of Horticulture
Pennsylvania State University
University Park, Pennsylvania

Chee-Kok Chin, Ph.D.
Associate Professor
Department of Horticulture and
  Forestry
Cook College
Rutgers University
New Brunswick, New Jersey

John J. Gaynor, Ph.D.
Assistant Professor of Plant Molecular
  Biology
Department of Botany
Rutgers University
Newark, New Jersey

Ravinder Kaur-Sawhney, Ph.D.
Research Associate
Department of Biology
Yale University
New Haven, Connecticut

C. R. Somerville, Ph.D.
Associate Professor
Department of Botany and Plant
  Pathology
MSU-DOE Plant Research Laboratory
Michigan State University
East Lansing, Michigan

S. C. Somerville, Ph.D.
Assistant Professor
MSU-DOE Plant Research Laboratory
Michigan Plant University
East Lansing, Michigan

Barbara A. Zilinskas, Ph.D.
Associate Professor of Plant
  Physiology
Department of Biochemistry and
  Microbiology
New Jersey Agricultural Experiment
  Station
Cook College
Rutgers University
New Brunswick, New Jersey

TABLE OF CONTENTS

Part I
Introductory Chapters

Chapter 1

PLANT BREEDING: BIOCHEMISTRY AND CROP PRODUCTIVITY

Carlos A. Neyra

TABLE OF CONTENTS

I. INTRODUCTION

One of the major challenges for scientists, at present, is to help increase effectively world food production to meet the ever increasing demands of a rapidly expanding population. A projection of current statistical figures indicates that world population will reach 8 billion by the first decade of the 21st century.[1] By that time, food supply will still be dependent, for the most part, on agricultural farming, but the efficiency of utilization of cultivated land will have to be greatly improved by the adoption of better farming practices and the development of new genotypes capable of achieving higher yields than those currently under cultivation.

The significant growth of agricultural productivity over the past few decades can be attributed in large part to the application of modern concepts of genetics to the breeding of superior crops. Most of the gains have come from the exploitation of the genetic diversity present in species that have long been domesticated but so far, all genotypes available for farming have been developed by conventional methods based on phenotypic characters associated with crop yields. To some extent, the goal of increasing crop productivity and the rate of achievements depends on our ability to understand and manipulate the biochemical and genetic mechanisms controlling plant productivity. Several biochemical criteria have been proposed for the selection of superior cultivars and have been under evaluation for almost 25 years. Yet, they are still to be implemented into a successful plant breeding program of economic importance.

This book presents a comprehensive survey of progress and current knowledge of those biochemical processes with greater potential for the development of superior cultivars: photosynthesis, photorespiration, nitrate assimilation, biological nitrogen fixation, and starch and protein synthesis. Each chapter addresses the major features of the physiological processes involved: the indentification of rate-limiting steps; regulatory properties of key enzymes; and a discussion of alternatives to overcome the limitations for enhanced crop productivity.

The contents of this book are presented in two volumes: Carbon Nutrition (Volume I) and Nitrogen Nutrition (Volume II). Throughout the contents, we have attempted to provide sufficient background information and discussion of biochemical tools to strengthen the approach and goals of plant breeding programs. Special attention was given to analyze the possibility of using genetic engineering methods for raising the genetic yield potential of crops. Modern plant breeding, in the context of this book, is seen as a combination of both the conventional or traditional plant breeding approach with the more novel molecular and cellularly based genetic engineering methodologies.

Because of the vast knowledge needed to make an in-depth presentation of the subjects under consideration, I chose to invite some of the best regarded scientists to contribute with their expertise to this book. As the editor, I was also responsible for the selection of the topics to be included. Within broad editorial guidelines, the contributors have been responsible for the precise content and approach of their chapters. A list of contributors and their institutional affiliations has been included for anyone willing to make further contacts and expand on the topics covered under this book.

II. PLANT BREEDING AND CROP PRODUCTIVITY

A. Food Crops and Partitioning of Assimilates

The welfare of mankind largely depends upon plants as a source of food and fiber. Of all plants known to have economical value (about 1500 species), only a few (around 30 species) have been developed as food crops.[1-3] One half of those species (eight cereals and seven legumes) is known as grain crops (pulses in the case of legumes), and these provide most of the calories and protein for human nutrition. In fact, 50% of those requirements is met by only three cereal crops: wheat, rice, and corn. Among the pulses, soybeans (*Glycine max*) and drybeans (*Phaseolus vulgaris*) are the most important food crops. Potatoes (*Solanum tuberosum*) and cassava (*Manihot esculenta*) are the most important among the crops harvested for their underground parts.[4] All of the crops cited above have been developed because of their capability to mobilize and accumulate in defined structures of the plant a large proportion of their biosynthetic production. Thus, the increase of the genetic yield potential, within a given crop species, has usually been accompanied by an increase in harvest index (dry matter accumulated in storage organs/total dry matter produced by the plant).[1,5-9]

The partitioning of assimilates among various plant parts is of major importance for agricultural productivity. We also know that an important variation exists in the relative allocation of dry matter to different organs. Crops grown for their vegetative aerial portion tend to reinvest more into the development of new leaves. Grain crops, on the other hand, tend to allocate more assimilates into the production of reproductive structures, and other crops (potato, cassava, etc.) invest relatively more into the underground plant parts. However, we know relatively very little about the physiological and biochemical mechanisms controlling a defined pattern of assimilate partitioning. Research in these areas should help provide the knowledge needed to devise modern plant breeding strategies. Further understanding of the specific nature of assimilate partitioning may lead to further manipulation of harvest indexes and maximization of reproductive yields in grain crops and vegetative yields in forage and root crops.

In principle, assimilate partitioning is based upon the relative strength of source and sink structures. Source tissues are those capable of exporting assimilates. Sinks, on the other hand, are represented by structures that exhibit a net import of assimilates, and their activity will be dependent upon size, position in relation to the source, geometry, physiological stage, and competition among different sinks.[5,9] All actively growing, storing, or metabolizing tissues may be considered as strong sinks. Mobilization of assimilates from source to sink occurs primarily via phloem.[9,10] Sucrose (and, to a lesser extent, amino acids) is universally recognized as the main form of carbon transported. Organic nitrogen, on the other hand, is primarily transported in the form of amino acids.[9,10] The interrelationships between primary carbon assimilation, and sucrose and starch metabolism are schematically presented in Figure 1 and will be discussed in more detail in Chapters 6 and 8 (Volume I). The interrelationships between the primary assimilation of inorganic nitrogen, temporary storage of organic nitrogen in vegetative structures, and the remobilization and transport of organic nitrogen to the seeds are also shown in Figure 1. For further details, see Chapters 4 and 5 (Volume II) and also Pate,[10] Below et al.,[11] Miflin and Lea,[12] and Reed et al.[13]

B. Wheat, Rice, and Corn Breeding

Plant breeders have been successful in developing cultivars that allocate a larger proportion of their assimilates to the harvestable organ without increasing the total biomass production (dry matter/ha), and this suggested that a genetic basis for partitioning exists.[1,5,9] In wheat and rice, the release of semi-dwarf varieties with an improved plant architecture led to higher harvest indexes with a larger fraction of the assimilates being partitioned into the grains.[1] The development of semi-dwarf varieties of wheat and rice, by incorporation of dwarfing genes through conventional breeding practices, marked the beginning of a new era in agriculture: the green revolution of the 1960s and 1970s, but required, in addition to the new varieties, the development of "technological packages" that included an increased use of fertilizers, irrigation, and chemical pesticides to allow maximum yield potentials to be expressed.

Another significant contribution made to agriculture by conventional breeding was the commercial introduction of the first corn hybrids in the U.S. about 40 years ago.[1]

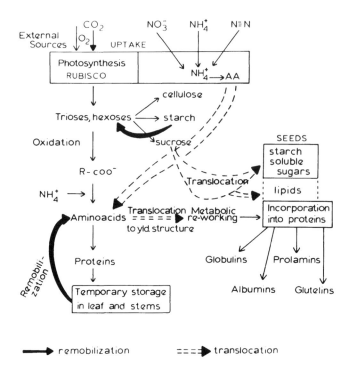

FIGURE 1.   Schematic illustration of the conversion of inorganic carbon and nitrogen into plant organic constituents. This scheme also illustrates the remobilization of carbon and nitrogen from vegetative to reproductive structures. Box on the top left represents a chloroplast where CO_2 fixation takes place with the participation of ribulose bisphosphate carboxylase. Box on the top right represents a chloroplast or a nodule for the assimilation of nitrate (NO_3), or atmospheric nitrogen (N_2) into amino acids, respectively.

Since then, many improved elite hybrids with continuously higher yields, disease and insect resistance, and shorter and stronger stalks suitable for mechanical harvesting have been developed. In spite of the fact that hybrid corn has been predominantly cropped in the last 25 to 30 years in the U.S., a marked increase in yields was further associated with the use of progressively (year after year) larger quantities of fertilizer N.[6,14]

The development of semi-dwarf wheat and rice genotypes together with the advent of corn hybrids is a clear example of successful plant breeding aimed at increasing crop productivity. Nonetheless, it should be pointed out that yield increases in all of those three cases were highly dependent on the availability and responsiveness to fertilizer nitrogen.[14] It is not surprising then that 50% of the total fertilizer nitrogen used in agriculture goes to cereal production.[15] However, economic considerations and protection of environmental quality would eventually limit the amounts of fertilizer nitrogen to be used in farming. The ever increasing prices of fertilizer nitrogen and the need of fossil fuels for its manufacture make it desirable to find alternative ways to lessen the dependence on high nitrogen inputs for obtaining maximal yields.[15] Genotypic differences in nitrogen utilization have been demonstrated, indicating that there is potential for developing superior, nitrogen-efficient genotypes.[16-20] Later, in Volume II (Chapter 4), a method will be discussed that uses dry matter and total N measurements to describe some of the traits which may indicate the effectiveness with which nitrogen assimilation processes contribute to dry matter or grain production and how these methods could help facilitate genetic selection of specific genotypes with improved efficiency for the utilization of available nitrogen.

C. Grain Legumes

Grain legumes represent some of the earliest agricultural crops cultivated for direct consumption of their seeds but currently they amount to only 10% of the total world production of cereal grains.[15] Even though legumes and cereals complement each other nutritionally and agriculturally in terms of cropping patterns, their progress has not run in parallel in recent years with cereal production increasing at a much faster rate than that of legumes.[14,15] The increased use of fertilizer N, together with the development of genotypes highly responsive to added N, is given as the main reason for the steady increase in cereal grain yields over the last 25 to 30 years.[14,15] Obviously, grain legumes have not been selected for their responsiveness to fertilizer N and many attempts to supplement nitrogen fixation with fertilizer N have not succeeded in increasing yields.[14,15,21,22,23] Nevertheless, there are reports indicating that grain legumes may still benefit from adequate timing and quantity of fertilizer N applied[24,25] and that the lack of appropriate agronomical practices to use fertilizer N in legumes may have prevented its use on a larger scale.

Geneticists and plant breeders have been effective in the use of existing genetic variability for the improvement of crop yields, but symbiosis and N_2-fixation characteristics have rarely been considered in plant breeding programs. The potential capability of legumes for nitrogen self-sufficiency should be pursued to its maximum, but yields have to be enhanced or at least maintained. Increasing N input for increasing grain legume yields will place a larger demand on photosynthate to satisfy the energy required for the reduction of nitrate (NO_3 assimilation) or atmospheric nitrogen (N_2-fixation). Thus, legume breeding programs should consider the increase of photosynthetic productivity simultaneously with the enhancement of N incorporation by these crops. For more detailed discussions on these topics, please refer to Chapters 2 and 6 (Volume II).

D. Root and Tuber Crops

On a world-wide basis, roots and tubers represent nearly half the volume of all horticultural crops and over 30% of the staple grains.[26] Potatoes are grown mainly in cool climates (highland tropics or northern and southern latitude temperate climates) and are considered the world's fourth most important food crop after rice, wheat, and corn.[27] Cassava, on the other hand, is a basic staple in the tropics and the leading horticultural crop in the developing world.[26,28] Cassava is regarded as one of the most efficient crops under water and nutrient stress conditions. Under optimum conditions, yields of over 80 tons of fresh roots per hectare per year (equivalent to ∿30 tons of dry matter) have been reported.[28] Under nutrient or water stress conditions, yields could be significantly reduced to 9 or 10 tons of fresh roots per hectare per year but still outyields most crops under such limiting conditions. The yield potential of cassava under stress conditions is a consequence of several physiological adaptive features of the plant. Cassava has a remarkably high harvest index of about 0.6 to 0.8 (ratio of root weight to total biomass production) even under the influence of stress factors. In fact, under nutrient or water stress conditions, cassava increases the relative allocation of dry matter to the roots.[28] Furthermore, during drought stress, cassava follows a conservative pattern of water use by reducing stomatal aperture and decreasing the formation of new leaves. Stomatal closure is quite sensitive to vapor pressure differences and they may respond even before bulk leaf water potential is affected.[29,30] Photosynthetic activity can be recovered immediately after rehydration. All of those adaptive characteristics, and their tolerance to acid-infertile soils with relatively high aluminum allow the use of cassava on the marginal lands of the tropics and subtropics to help increase total agricultural production.[28]

Currently cultivated cassava and potato cultivars have been developed from a rather narrow genetic base. Both of these crops originated in America but were broadly dispersed throughout the world. Large, untapped genetic resources still exist in the centers of origin but their potential value is, for the most part, unknown; however, they could serve as the basic genetic resources for wider climatic flexibility with higher yields and better resistance to pests and disease.[27] With the organization of the International Research Centers having specific mandates, it has been possible to broaden the genetic base of both potatoes and cassava. A world potato collection is being organized and maintained by the International Potato Center (CIP), headquartered at LaMolina, Lima, Peru. The size of the collection now stands at 5000 accessions of primitive cultivars and 1291 accessions of wild tuber-bearing species. On the other hand, the International Center of Tropical Agriculture (CIAT) has developed a germplasm bank for cassava containing about 2600 accessions obtained from 13 Latin American and 2 Asian countries.

Innovative in vitro methods have been integrated to breeding programs for the conservation and exchange of potato and cassava germplasm.[31,32] Because of their high genetic stability and disease-free conditions, meristem and shoot tip cultures have become increasingly important as a means of propagation, conservation, and germplasm exchange. The in vitro system has been well accepted by several countries as a means to improve the phytosanitary aspects of germplasm exchange.[31,32] Meristems, isolated cells, and protoplast cultures have all been shown to regenerate normal plantlets and may add a new dimension to the breeding efforts in root and tuber crops. For further information on these topics, please refer to Chapters 2 and 3 of this book (Volume I).

III. PLANT BIOCHEMISTRY AND BREEDING RESEARCH

A. An Overview

Man has come a long way since we learned, about 10,000 years ago, how to cultivate plants. Until late in the 19th century, however, crop improvement was in the hands of farmers who selected the seed from preferred plant types or populations for subsequent sowing.[1,2,33] Genetics, on the other hand, was only known as a science at the start of the 20th century, after the laws of heredity proposed by Gregor Mendel were finally accepted.[1,34] Since genetics is basic to rational plant breeding, we should consider the practice of controlled breeding a rather new happening. Even though plant breeding is considered both an art and a science,[2,7] it should be remembered that breeding is central to the problem of enhancing crop productivity and must be scientifically based.[7] At least two scientific disciplines, genetics and biochemistry, are expected to provide the supporting basis to plant breeding research for the development of improved cultivars.

In the process of cultivar selection, traditional breeders have generally made use of phenotypic characters in their search for higher yielding varieties.[2,7] However, one would expect that in this process of selection, biochemical and physiological modifications have taken place. With the advent of new biochemical methods, plant physiologists and biochemists have been able to elucidate and interpret a large number of biochemical reactions associated with plant function, rate-limiting steps in pathways, and regulatory properties of the enzymes involved.

The detailed description of biochemical pathways and regulatory mechanisms of key enzymes in microorganisms have greatly contributed to the development of highly efficient and profitable industrial processes.[35-37] Advanced research into the biochemistry and genetics of the microorganisms involved has been the key to the success of industrial microbiology. On the other hand, the use of biochemical criteria for the development of superior plant genotypes is not a new idea,[38,39] but these genotypes have not been used directly, in any successful plant breeding program of economical importance, including the genotypes developed during the "green revolution" years of the 1960s and 1970s. Nonetheless, substantial progress has been made in the last two decades towards our understanding of plant biochemical reactions, most noticeably in the fields of carbon and nitrogen metabolism. The current knowledge on these two fields and its applicability to future plant breeding programs will be dealt with in more detail throughout this book.

Plants are largely dependent on solar energy, carbon, and nitrogen for maximal productivity. The production of dry matter on earth is estimated to be around 200 billion metric tons per year. This would require the reduction, using part of the solar energy fixed, of 70 billion and 0.2 billion tons of carbon and nitrogen, respectively. Since carbon is obtained from a highly oxidized atmosphere (20% O_2) and the concentration of CO_2 is relatively low (0.03%), the interactions between photosynthesis and photorespiration are of major concern for the purpose of maximizing yields (see Chapter 7, Volume I) in addition to the relationships between rates, duration of photosynthesis, and yield output.[40] Solar radiation, on the other hand, is the primary source of energy for the assimilation of both carbon and nitrogen,[41] but the efficiency of conversion of light energy into plant products is still very low. Since energy is lost in diverse ways between the arrival of sunlight at canopy level and the reduction of CO_2 to carbohydrate in the leaf,[40] more must be learned about the efficiency of light energy capture and conversion to make a fuller use of the energy available for crops. The potential for improvement of crop yields through changes in the efficiency and/or capacity of the light-dependent photosynthetic reactions is discussed in Chapter 4 (Volume I).

On a global scale, it has been estimated that biological nitrogen fixation contributes about 100×10^6 tons of nitrogen fixed per annum in agricultural soils.[15,27,30] Chemically produced fertilizer N contributes an additional 50×10^6 tons of nitrogen to agriculture.[15] These figures speak for themselves about the importance of both nitrogen sources (biologically and chemically fixed) for plant growth and agricultural productivity. Nonetheless, the efficiency of utilization of those N sources will depend upon the rate and duration of N uptake (either from soil or atmospheric sources), translocation, transient accumulation in vegetative organs, redistribution and accumulation of N in harvestable organs, sink size for N, etc. A broader discussion on all of these parameters will be found in Volume II of this book which is devoted entirely to nitrogen nutrition.

B. Gene Mutations and Cellular Approaches to Breeding

The primary goal of selection in plant breeding is the identification of desirable genotypes and effective selection is dependent upon the existence of genetic variability.[42] The extent of genetic variability in a specific breeding population depends on the germplasm included in it and its selection history. When genetic variability in a breeding program is insufficient to permit attainment of a specific goal, it will be necessary to increase the variability by using either mutagenic treatments or introduction of new germplasm.[42]

Gene mutation provides the basic material for natural selection.[2,34,43] The use of strains carrying a mutant gene can be useful probes in physiological investigations (see Chapters 4, 7, and 8, Volume I), or the mutation may carry a desirable characteristic that may be important to perpetuate.[39,44] The identification of specific mutants of agronomically important genes and regulation of their expression is also important to expand our understanding of plant molecular biology as the basis for future applications to the development of new genotypes by genetic engineering.[39,44,45] Gene mutation may be assumed to have proceeded at a more or less steady rate, age after age, throughout evolution.[2] The frequency of occurrence of spontaneous mutants is, however, very low, and special screening procedures should be adopted for their identification.[46] Alternatively, specific mutations may be induced by a number of chemical and physical treatments.[47] The latter have been very useful and extensively applied to microbial biochemistry research, but induction of stable mutations in higher plants presents a number of practical and methodological problems. However, the recent advances in plant cell and tissue culture techniques have set the stage for genetic manipulations of higher plant cells in much the same manner as microbial populations.[39,47] As plant cells are totipotent, for some species one could obtain adult plants from callus tissue; from this emerges the outline of new systems for investigating plant genetics and breeding for the development of superior plant genotypes. First, individual cells can be isolated from plant tissues and grown in a defined culture media. A large population of cells may then be available in a matter of days or weeks, depending on the type of tissue used as a source material and the plant of origin. Second, in such cell cultures, the rate of mutation may be increased, making possible the needed genetic variability and diversity of phenotypes from which to select. Third, foreign genetic material may be introduced directly into plant cells by following DNA-manipulation procedures.

A major limitation of phenotypic selection in vitro is that it may identify modifications of traits that are expressed at the cellular level only. The developmental complexity of higher plants imposes a severe limitation on the number of possible applications of the new methodologies based on cell or tissue culture techniques. Many traits of agronomic importance are the products of the organization of highly differentiated cells and, therefore, they may not appear in culture.[37,45,48] All of these limitations, however, may be resolved with the progress of research, and we have a better understanding and ability to manipulate the molecular and biochemical principles controlling correlative functions in differentiated organisms. The techniques available for the culture of cells and plantlets in vitro will certainly improve rapidly in the next few years and the use of in vitro selections applicable to the improvement of crop plants may be a reality sooner than expected (see Chapters 2 and 3, Volume I). Modern plant breeding may then be seen as a combination of both the conventional or traditional plant breeding approach with the more novel, molecular and cellularly based genetic engineering methodologies. Integration of these two approaches will certainly benefit science and agricultural productivity. This new era of research must rely strongly on the progress made in plant biochemistry research and its extension to modern plant breeding and the goal of improving the genetic potential of crops.

C. Protein Quantity and Quality

The explosive yield increases of cereal grains in recent decades are the best indicator of how successfully genetic resources can be exploited.[49] Because the percent protein content of grains has remained virtually unchanged over the last 30 years,[14] we could deduce that the total quantity of crude protein per unit of land has increased in proportion to the increase in grain yields. Quantitatively, it is estimated that cereal grains provide about 50% of the world protein needs; grain legumes and animal products, 20 and 30%, respectively.[50]

Grain protein concentration can be influenced by heredity as well as cultural conditions.[14,49,51] Genetic constitution is, however, the most important single factor affecting protein levels in grains.[51] There is ample evidence showing that substantial genetic variation exists for grain protein concentration.[49,51] Even though it would be highly desirable to increase the protein concentration in cereal grains, yields and desirable commercial characteristics should be maintained. Frey[52] reported that in various cereal species the protein concentration and grain yields are negatively correlated, with some exceptions.[53] Also, Lambert et al.[54] found that opaque-2 corn hybrids had higher protein concentrations but lower yields than their normal counterparts. Other studies have shown similar inverse relationships in corn hybrids.[48] However, wheat genes from "Atlas 66" have been utilized to elevate protein concentration in "Lancota" cultivar by 1 to 2% without a reduction in grain yields.[49,53] Similarly, genes for high yield that have no effect on groat protein have been discovered in oats.[48] Thus, the possibility exists for increasing protein concentration without affecting yields of cereal grains.

Further consideration should also be given to improve the nutritional quality and amino acid balances in both cereal and legume grains. We know that grain yields are positively correlated with the amount of N reaching to the grains and plant breeders have succeeded in providing genotypes with a high response to fertilizer N.[1-14] We also know that part of the N reaching to the grains is accumulated into various types of storage proteins (see Figure 1). The major classes of storage proteins have been classified according to their solubility properties into albumins (water soluble), globulins (soluble in salt), prolamins (soluble in strong alcohol), and glutelins (soluble in dilute alkali solutions).[49,55] This classification has been very valuable in grain protein research for the improvement of their nutritional value.

Most grain cereals contain primarily prolamine type proteins while globulins are the main storage type proteins in legume grains. Prolamines from all cereals are rich in proline and glutamine but deficient in lysine and tryptophan.[49] Globulins, on the other hand, are rich in amides but low in sulfur-containing amino acids.[49] Thus, the protein quality of grains is determined by the relative proportions of the major storage proteins and one approach to improve the nutritional quality of grains would be to alter the balance of storage protein types. Mertz et al.[56] found that the opaque-2 mutant of maize shows significantly higher lysine and tryptophan than normal maize. The opaque-2 gene drastically reduces the proportion of zein (prolamine type) and increases the proportions of water and salt-soluble proteins.[49] Similar gross effects have been found in barley and sorghum.[49,51]

Genetic engineering of seed storage proteins offers an alternative approach to improve the nutritional quality of grains.[37] Thus, the coding capacity of the genes could be altered to engineer specific storage proteins with a better amino acid balance[37] but it still remains to be seen how easily the genetic engineering of protein quality will also meet grain yields, protein contents, and market quality requirements for a given crop. Although all proteins may potentially function as an N sink, the primary functional role is, in all likelihood, played by the major storage protein(s) and their synthesis. In maize kernels, for example, zein is the main single storage protein accumulated and its synthesis can be readily manipulated by N fertilization as well as genetic means. Increase in kernel weight and yield has been shown to be positively correlated with increases in zein content.[57,58] Increases in N deposition in the normal endosperm induced by N fertilizer are confined primarily to zein.[57,58] Early termination of zein accumulation in the opaque-2 mutant of maize results in a reduction of sucrose movement into the kernels.[57] Thus, the much reduced zein content of the opaque-2 mutant may be responsible for the observed lower yields as compared to their normal counterparts.

ACKNOWLEDGMENTS

New Jersey Agricultural Experiment Station, Publication No. F-01103-1-84, supported in part by State funds and by the United States Hatch Act is acknowledged. Partial support was also provided by a Rutgers Research Council Grant. Thanks are extended to Mrs. Alice Montana for the typing of this manuscript, to Sylvia Taylor for skillful assistance in the preparation of the illustrations, and to Dr. Barbara A. Zilinskas for reading and helpful comments on this manuscript.

REFERENCES

1. Bourlaug, N. E., Contributions of conventional plant breeding to food production, *Science*, 219, 689, 1983.
2. Lawrence, W. J. C., *Plant Breeding,* Studies in Biology Series No. 12, Edward Arnold, London, 1968.
3. Vietmeyer, N., The greening of the future, *Quest,* (Sept.), 26, 1979.
4. Burton, W. G. and Booth, R. H., Post-harvest technology for developing country tropical climates, in *Proc. Int. Congr. Res. for the Potato in the Year 2,000,* Hooker, W. J., Ed., International Potato Center (CIP), Lima, Peru, 1983, 40.
5. Evans, L. T., Raising the yield potential: by selection or design, in *Genetic Engineering of Plants. An Agricultural Perspective,* Kosuge, T., Meredith, C. P., and Hollander, A., Eds., Plenum Press, New York, 1983, 371.
6. Hardy, R. W. F., Havelka, W. D., and Quevedaux, B., Increasing crop productivity: the problem, strategies, approach, and selected rate limitations related to photosynthesis, in *Proc. 4th Int. Congr. on Photosynthesis,* Great Britain, 1977, 695.
7. Simmonds, N. W., Plant breeding: the state of the art, in *Genetic Engineering of Plants. An Agricultural Perspective,* Kosuge, T., Meredith, C. P., and Hollander, A., Eds., Plenum Press, New York, 1983, 5.
8. Zelitch, I., Photosynthesis and plant productivity, *Chem. Eng. News,* (Feb.), 28, 1979.
9. Hanson, A., Plant breeding and partitioning in cereals and grain legumes, in *Workshop on Partitioning of Assimilates,* Michigan State University, American Society of Plant Physiologists, Rockville, Md., 1979, 18.
10. Pate, J. S., Transport and partitioning of nitrogenous solutes, *Annu. Rev. Plant Physiol.,* 31, 313, 1980.
11. Below, F. E., Christensen, L. E., Reed, A. J., and Hageman, R. H., Availability of reduced N and carbohydrates for ear development of maize, *Plant Physiol.,* 68, 1186, 1981.
12. Miflin, B. J. and Lea, P. J., Amino acid metabolism, *Annu. Rev. Plant Physiol.,* 28, 299, 1977.
13. Reed, A. J., Below, F. E., and Hageman, R. H., Grain protein accumulation and the relationship between leaf nitrate reductase and protease activities during grain development in maize (*Zea mays* L.). I. Variation between genotypes, *Plant Physiol.,* 66, 164, 1980.
14. Hageman, R. H., Integration of nitrogen assimilation in relation to yield, in *Nitrogen Assimilation of Plants,* Hewitt, E. J. and Cutting, C. V., Eds., Academic Press, New York, 1979, 591.
15. Hardy, R. W. F. and Havelka, W. D., Nitrogen fixation research: key to world food?, in *Food: Politics, Economics, Nutrition and Research,* Abelson, P. H., Ed., Science, (Spec. Compendium), 1975, 178.
16. Beauchamp, E. G., Kannenberg, L. W., and Hunter, R. B., Nitrogen accumulation and translocation in corn genotypes following silking, *Agron. J.,* 68, 418, 1976.
17. Chevalier, P. and Schrader, L. E., Genotypic differences in nitrate absorption and partitioning of nitrogen among plant parts, *Crop. Sci.,* 17, 987, 1977.
18. Moll, R. H. and Kamprath, E. J., Effects of population density upon agronomic traits associated with genetic increases in yield of *Zea mays* L., *Agron. J.,* 69, 81, 1977.
19. Moll, R. H., Kamprath, E. J., and Jackson, W. A., Analysis and interpretation of factors which contribute to efficiency of nitrogen utilization, *Agron. J.,* 74, 562, 1982.
20. Reed, A. J. and Hageman, R. H., Relationship between nitrate uptake, flux and reduction and the accumulation of reduced nitrogen in maize (*Zea mays* L.). I. Genotypic variation, *Plant Physiol.,* 66, 1179, 1980.
21. Franco, A. A., Nutritional restraints for tropical grain legume symbiosis, in *Exploiting the Legume — Rhizobium Symbiosis in Tropical Agriculture,* (Misc. 145), Vincent, J. M., Whitney, A. S., and Bose, J., Eds., College of Tropical Agriculture, University of Hawaii, Honolulu, 1977, 237.
22. Gibson, A. H., Limitation to dinitrogen fixation by legumes, in *Proc. 1st Int. Symp. Nitrogen Fixation,* Newton, W. E. and Nyman, C. J., Eds., Washington State University Press, Pullman, Wash., 1976, 400.
23. Havelka, U. D., and Hardy, R. W. F., Legume N₂ fixation as a problem in carbon nutrition, in *Proc. 1st Int. Symp. Nitrogen Fixation,* Washington State University Press, Pullman, Wash., 1976, 456.
24. Franco, A. A., Pereira, J. C., and Neyra, C. A., Seasonal patterns of nitrate reductase and nitrogenase activities in *Phaseolus vulgaris* L., *Plant Physiol.,* 63, 421, 1979.
25. Garcia, L. R. and Hanway, J. J., Foliar fertilization of soybeans during the seed filling period, *Agron. J.,* 68, 653, 1976.
26. Burton, W. G. and Booth, R. H., Post-harvest technology for developing country tropical climates, in *Research for the Potato in the Year 2000,* Hooker, W. J., Ed., International Potato Center, Lima, Peru, 1983, 41.

27. Swaminathan, M. S. and Sawyer, R. L., The potential of the potato as a world food, in *Research for the Potato in the Year 2000,* Hooker, W. J., Ed., International Potato Center, Lima, Peru, 1983, 3.

28. Cock, J. H., Cassava: a basic energy source in the tropics, *Science,* 218, 755, 1982.

29. El-Sharkawy, M. A. and Cock, J. H., Water use efficiency of cassava. I. Effects of air humidity and water stress on stomatal conductance and gas exchange, *Crop Sci.,* 24, 497, 1984.

30. El-Sharkawy, M. A., Cock, J. H., and Held, A., Water use efficiency of cassava. II. Differing sensitivity of stomata to air humidity in cassava and other warm-climate species, *Crop Sci.,* 24, 503, 1984.

31. Roca, W. M., Espinoza, N. O., Roca, M. R., and Bryan, J. R., A tissue culture method for the rapid propagation of potatoes, *Am. Potato J.,* 55, 691, 1978.

32. Roca, W. M., Rodriguez, J., Beltran, J., Roa, J., and Mafla, G., Tissue culture for the conservation and international exchange of germplasm, in *Proc. 5th Int. Congr. on Plant Tissue and Cell Culture,* Fujiwara, A., Ed., Plant Tissue Culture, 771, 1982.

33. Turnham, D., Sources of agricultural growth, in *World Development Report,* Oxford University Press, 1982, 57.

34. Herskowitz, I. H., *Principles of Genetics,* Macmillan, New York, 1973.

35. Wallace, D. H., Ozbun, J. L., and Munger, H. M., Physiological genetics of crop yield, *Adv. Agron.,* 24, 97, 1972.

36. Carlson, P. S. and Hoisington, C. G., The new genetics: notes from Demeter's workshop, in *The Antioch Reviews,* Fogarti, R. S., Ed., Antioch Review Inc., Yellow Springs, Ohio, 1980, 409.

37. Larkins, B. A., Genetic engineering of seed storage proteins, in *Genetic Engineering of Plants, An Agricultural Perspective,* Kosuge, T., Meredith, C. P., and Hollander, A., Eds., Plenum Press, New York, 1983, 93.

38. Hageman, R. H., Leng, E. R., and Dudley, J. W., A biochemical approach to corn breeding, *Adv. Agron.,* 19, 45, 1967.

39. Nelson, O. E. and Burr, B., Biochemical genetics of higher plants, *Annu. Rev. Plant. Physiol.,* 24, 493, 1973.

40. Ogren, W. L., Increasing carbon fixation by crop plants, in *Proc. 4th Int. Congr. on Photosynthesis,* Great Britain, 1977, 721.

41. Neyra, C. A., Interactions of plant photosynthesis with dinitrogen fixation and nitrate assimilation, in *Limitations and Potentials of Nitrogen Fixation in the Tropics,* Vol. 10, Hollander, A., Ed., Basic Life Sciences, Plenum Press, New York, 1978, 111.

42. Hallauer, A. R., Selection and breeding methods, in *Plant Breeding II,* Frey, K. J., Ed., Iowa State University Press, Ames, 1981, 3.

43. Briggs, F. N. and Knowles, P. F., *Introduction to Plant Breeding,* Reinhold Publishing, New York, 1967, 301.

44. Levine, R. P., The analysis of photosynthesis using mutant strains of algae and higher plants, *Annu. Rev. Plant. Physiol.,* 20, 523, 1969.

45. Miflin, B. J., Bright, S. W. J., Rognew, S. E., and Kuck, J. S. H., Amino acids, nutrition and stress: the role of biochemical mutants in solving problems of crop quality, in *Genetic Engineering of Plants. An Agricultural Perspective,* Kosuge, T., Meredith, C. P., and Hollander, A., Eds., Plenum Press, New York, 1983, 391.

46. Burton, G. S., Meeting human needs through plant breeding: past progress and prospects for the future, in *Plant Breeding II,* Frey, K. J., Ed., Iowa State University Press, Ames, 1981, 433.

47. Coe, E. H., Jr. and Neuffer, M. G., The genetics of corn, in *Corn and Corn Improvement,* No. 18, Sprague, G. F., Ed., American Society of Agronomy, Madison, Wisc., 18, 1977, 111.

48. Chaleff, R. S., Isolation of agronomically useful mutants from plant cell cultures, *Science,* 219, 676, 1983.

49. Axtell, J. D., Breeding for improved nutritional quality, in *Plant Breeding II,* Frey, K. J., Ed., Iowa State University Press, Ames, 1981, 365.

50. Oram, R. N. and Brock, R. D., Prospects for improving protein yield and quality by breeding, *J. Aust. Inst. Agric. Sci.,* 38, 163, 1972.

51. Alexander, D. E. and Creech, R. G., Breeding special industrial and nutritional types, in *Corn and Corn Improvement,* No. 18, Sprague, G. F., Ed., American Society of Agronomy, Madison, Wisc., 1977, 363.

52. Frey, K. J., Proteins of oats, *Z. Planzenzüchtg.,* 78, 185, 1977.

53. Johnson, V. A., Mattern, P. J., and Kuhr, S. L., Genetic improvement of wheat protein, in Proc. Symp. on Seed Protein Improvement in Cereals and Grain Legumes, Neuherberg, West Germany, 1979, 165.

54. Lambert, R. J., Alexander, D. E., and Dudley, J. W., Relative performance of normal and modified protein (opaque-2) maize hybrids, *Crop Sci.,* 9, 242, 1969.

55. Osborne, T. B., The amount and properties of the proteids of the maize kernel, *J. Am. Chem. Soc.,* 19, 525, 1897.
56. Mertz, E. T., Bates, L. S., and Nelson, O. E., Mutant gene that changes protein composition and increases lysine content of maize endosperm, *Science,* 145, 279, 1964.
57. Tsai, C. Y., Huber, D. M., and Warren, H. L., A proposed role of Zein and Glutelin as N sinks in maize, *Plant Physiol.,* 66, 330, 1980.
58. Tsai, C. Y., Huber, D. M., Glover, D. V., and Warren, H. L., Relationship of N deposition to grain yield and N response of three maize hybrids, *Crop Sci.,* 24, 277, 1984.

Chapter 2

PRODUCTION OF NOVEL CROPS BY SOMATIC HYBRIDIZATION

John J. Gaynor and Ravindar Kaur-Sawhney

TABLE OF CONTENTS

I. INTRODUCTION

Traditional breeding programs have made great strides in recent years in improving both the environmental range and total yields for many important food crops. Two of the most striking examples are wheat and maize, which have experienced a doubling and tripling, respectively, in yields per hectare in the past 40 years.[1] In addition, rice and other agriculturally important crops have also witnessed substantial, although less dramatic, improvements in total yields.[2] However, despite these successes, traditional breeding procedures are plagued by several severe limitations.

A. Shortcomings of Traditional Breeding

One of the most serious limitations of traditional plant breeding is that of sexual incompatibility. Sexual incompatibility can occur between phylogenetically distant plants and even between closely related plants (as seen in the sexual incompatibility of several varieties of rice[2]). Generally speaking, the more distantly related the parents, the more difficult it is to produce a hybrid between them, due to the various hybridization barriers established during the course of evolution. Sexual compatibility may fail at any number of critical steps, namely, pollen germination, pollen tube growth, fertilization, or seed development.[3] A defect in any one or more of these steps will render two plants sexually incompatible. Although sexual incompatibility presents a formidable barrier to the plant breeder, it can be overcome by a process known as somatic hybridization.

II. SOMATIC HYBRIDIZATION

Somatic hybridization circumvents the problem of sexual incompatibility by artificially combining the genomes of two or more parents in a single hybrid. This is done by fusing together nonreproductive somatic (or "body") cells from one parent with those of another in vitro. This process is outlined in a general fashion in Figure 1.

A. Protoplast Isolation and Fusion

Leaf mesophyll protoplasts are commonly used as the starting material for somatic hybridization.[4] Briefly, protoplasts are produced by either peeling the epidermal layer from the leaf or by finely dicing the leaf and then exposing the mesophyll tissue to a combination of cellulase(s) and pectinase(s) which enzymatically dissolve the cell wall. The cells released from the leaf are spherical due to the fact that their supporting wall has been stripped away. Such isolated protoplasts represent viable plant cells which, in many cases, are capable of sustained cell division and eventually capable of regenerating whole plants.

Once isolated, protoplasts can be readily fused with other protoplasts. There are two procedures for inducing protoplast fusion: (1) polyethylene glycol (PEG)-induced fusion; and (2) electrofusion. In the former method, polyethylene glycol polymers $[HOCH_2-(CH_2-O-CH_2)_n-CH_2OH]$ are added (at approximately 40% w/v) to a mixture of plant protoplasts, which causes massive clumping of cells and subsequent fusion of some pairings. PEG-induced fusion rates vary from 1 to 20%, with a mean fusion rate of approximately 1 to 2%. Fusion is dependent upon close membrane contact and can be affected by pH, temperature, $[Ca^{++}]$, and purity of PEG, as well as other poorly defined factors. Despite the rather variable fusion rates and the low yield of fusion products, PEG remains the most widely used fusogen for the production of somatic hybrids.[5]

A recent alternative to PEG fusion is the fusion of plant protoplasts by an electric field, or simply, electrofusion. This technique was described independently by Senda

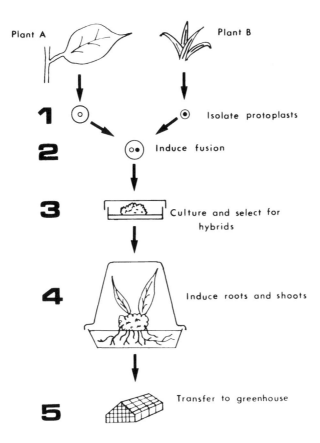

FIGURE 1.   General scheme for the production of somatic hybrids. See text for details.

et al.[6] and Zimmermann and colleagues.[7-9] Although the biophysics of this method are beyond the scope of this chapter (cf. reviews by Gaynor,[10] Zimmermann,[11] and Zimmermann and Vienken[12]), the technique can be outlined as follows. Electrofusion is a two-stage process: (1) the cells are first brought into intimate contact, between two electrodes, by the use of a high frequency AC field; and (2) the cells are fused by the application of a single high-voltage DC pulse of very short duration. Figure 2 illustrates a typical fusion sequence induced by this procedure.

In contrast to the PEG-induced fusion technique, electrofusion routinely obtains rates of 80% or better, does not subject the cells to abnormally high osmotic pressures, and does not appear to be damaging to most plant protoplasts.[13] The advantage to this method is that fusion is rapid (usually completed in less than 10 min), highly synchronous (all cells fuse at once), and offers exceptional control over both the ploidy level obtained (based on the numbers of cells held in a single pearl chain) and, more importantly, the types of cells which are fused. In addition, we have recently demonstrated that electrofusion is a relatively mild technique in that viable fusion products can be obtained which, in some cases, can be regenerated into whole plants.[13]

One point which should be emphasized, and is critical to the production of somatic hybrids, is the ability to control the fusion of different cell types. Obviously, in the production of somatic hybrids, one wants to maximize the number of heterokaryons (unlike pairings) and minimize the number of homokaryons (like pairings) that are produced. PEG fusion, since it is a random process, offers little or no control in this respect; electrofusion, on the other hand, since the pairing of different protoplast types

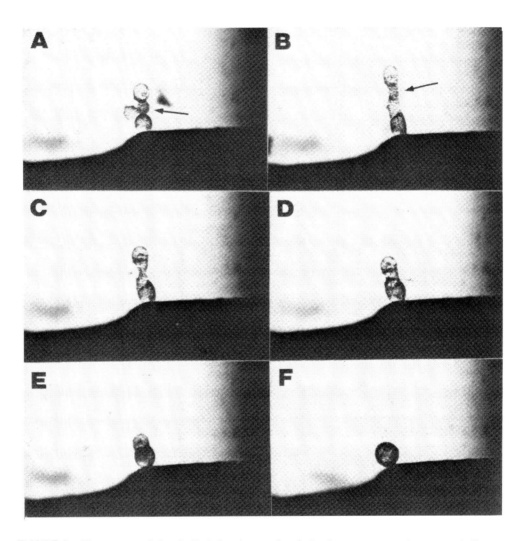

FIGURE 2.   Time-course of electrically induced protoplast fusion between corn and oat mesophyll protoplasts. Frame A shows the cells immediately before fusion is initiated (the three larger cells are oat mesophyll
protoplasts and the smaller, darker-stained protoplast is from corn mesophyll and is indicated by an arrow).
Frame B shows the cells 10 sec after the fusion pulse was applied. Note how the protoplast to one side has
"snapped" into line as a result of the fusion pulse. Cells have elongated and membrane interfaces are flattened
indicating that fusion has begun. Frame C (1 min), D (3 min), E (5 min), and F (14 min) represent successive
time intervals after fusion. Somatic hybrids are then removed from the electrode surface and cultured. Like
the two parental lines, *Zea-Avena* hybrids produced by this method have not divided.[31] See text for further
details.

can be controlled by varying the density and flow rates of cells as well as the strength
of the AC field, offers very fine control over this process. This concept is illustrated in
Figure 2. Here we see the electrically induced fusion of three oat protoplasts with a
single corn mesophyll protoplast, which is smaller and appears darker due to vital
staining with neutral red. This was accomplished by first passing purified oat protoplasts into the fusion chamber and aligning the cells in short chains along the electrode
surface. Next, purified corn mesophyll protoplasts were introduced to the same chamber (at very low densities) and these cells, because of the characteristics of the electric
field, were attracted to the oat protoplasts already sticking to the electrodes. Fusion
was then initiated by the application of a single high-voltage DC pulse, and within 15

min, fusion was complete. This illustrates that one can easily and quickly control the pairings of different cell types to yield unique somatic hybrids by the electrofusion method. Using this technique we have been able to produce *Zea-Avena, Avena-Amaranthus, Nicotiana-Petunia,* and *Avena-Petunia* heterokaryons. We have not, as yet, attempted to regenerate any of these somatic hybrids. In all probability, however, these would not have been viable since at least one of the parents in each case was incapable of regeneration.

B. Selection of Somatic Hybrids

Once somatic hybrids are produced by protoplast fusion, one must select for and culture such hybrids. As should be obvious from the above discussion on protoplast fusion, several different pairings are possible when two cell types are fused. The trick in somatic hybridization is to be able to select *for* novel heterokaryons resulting from the fusion of protoplast A with protoplast B and/or to select *against* A-A and B-B type pairings. Numerous selection schemes have been devised to select for such hybrids and are usually based on physical, genetic, morphological, or biochemical characteristics.[5] Figure 3 illustrates a selection technique which we have used to successfully enrich for heterokaryons from several different protoplast fusion experiments. This method exploits the density difference between two protoplast populations, in this case, *Amaranthus* and *Avena* protoplasts. *Avena* protoplasts were isolated by a standard procedure[13] and purified on a sucrose step gradient[14] identical to that in Figure 3 to yield a single band at a density of 1.058 gm/cc. Likewise, *Amaranthus* epidermal protoplasts were prepared as described previously[15] and isolated on a separate but identical sucrose gradient to yield a single band at a density of 1.030 gm/cc. In addition to there being a significant difference in density between these two protoplast types, they were also morphologically very distinct: the denser oat protoplasts were green due to the presence of mature chloroplasts whereas the lighter *Amaranthus* epidermal protoplasts were bright red due to the presence of betacyanin in the vacuole (also no chloroplasts were present in these cells). Thus, one would predict that when two such distinct cell types were mixed and fused, the resulting somatic hybrids could be identified by two independent criteria: (1) a density intermediate between the two parents; and (2) the presence of both red (betacyanin) and green (chloroplasts) colors in the same cell. When both cell types were mixed and subjected to electrofusion and then reisolated on identical density gradients, one finds, in addition to the two original parental bands (at a density of 1.030 and 1.058 gm/cc), a range of intermediate bands which also contain both chloroplasts (from oat protoplasts) and betacyanin (from *Amaranthus* protoplasts) as seen in Figure 3. If one could precisely regulate the fusion of oat to *Amaranthus* protoplasts in a 1:1 fashion, then only a single intermediate band should be found. However, we find experimentally that a range of intermediate density bands are produced suggesting a similar range in the numbers and types of cells fused originally. With the advent of procedures for artificially increasing the density of protoplasts (e.g., by evacuolation), we suggest that a method such as that presented here might serve as a quick and easy way of enriching for somatic hybrids.

In addition to density gradient enrichment, many other selection procedures have been used quite successfully. These are usually based on the expression of known genetic markers, physiological growth requirements, or by microisolation.[5]

After successful fusion and selection for somatic hybrids, the resulting material is cultured as callus and, in many cases, can be induced to produce both shoots and roots. In general, high cytokinin and low auxin induces shoots whereas low auxin and no cytokinin induces root formation. When such hybrids are sufficiently mature, they can be slowly weaned off of exogenous carbon sources and transferred to the greenhouse and, eventually, to the field.

Heterokaryon Enrichment
by Density Gradient Centrifugation

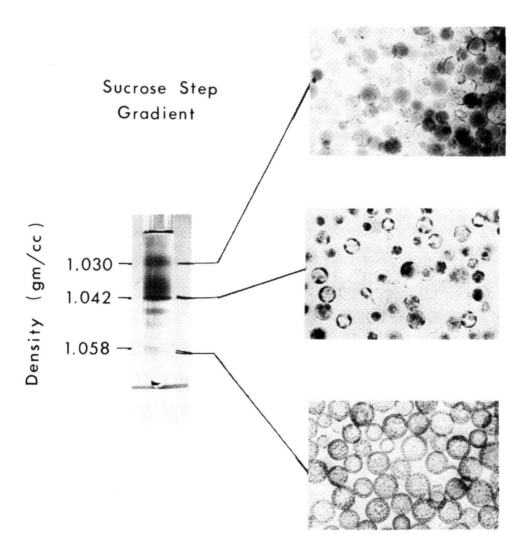

FIGURE 3.   Enrichment for somatic heterokaryons by density gradient centrifugation. If the two parental protoplast lines differ significantly in density, or can be made to be different, then selecting for heterokaryon fusion products is a simple matter of separating protoplasts over a density gradient. In this example, *Amaranthus* epidermal protoplasts, which are colored bright red because of the betacyanin in their vacuoles, were isolated and purified on a sucrose step gradient[14] similar to that shown here yielding a single band with a density of 1.030 gm/cc. Likewise, green oat mesophyll protoplasts were isolated and purified over a similar sucrose gradient yielding a single band at a density of 1.058 gm/cc. When these two protoplast types are mixed and fused by electrical means, the heterokaryon fusion products are isolated by subjecting the mixture to density gradient centrifugation over a sucrose step gradient as seen in this figure. Here one sees the original two parental cell types at their respective densities and also a range of intermediate density cells which represent heterokaryon fusion products (*"Amavena"* or *"Avaranthus"*), which are colored both red and green, of varying degrees of ploidy.

Table 1

INTERGENERIC SOMATIC HYBRID PLANTS PRODUCED BY
PROTOPLAST FUSION

Parent		Regenerated hybrid[a]		
		Asymmetric	Symmetric	
A	B	A B		Ref.
Arabidopsis thaliana	*Brassica campestris*	+ +		16, 17
Datura innoxia	*Atropa belladonna*	+		18
Daucus carota	*Petroselinium hortense*	+		19
	Aegopodium podagraria	+		20
Lycopersicon esculentum	*Solanum tuberosum*	+ +		21, 22
Nicotiana tabacum	*Petunia hybrida*	+		23
	S. sucrense	+		24
	S. tuberosum	+		24
	Salpiglossis sinuata		+	25

[a]   A (+) indicates whether or not the hybrid regenerated from the indicated fusion is sym-
metric (like both parents) or asymmetric (more like one parent). A (+) under just A or
B indicates that the hybrid resembles that parent; a (+) under both parents means that
asymmetric hybrids have been produced which resemble both parents A and B with
equal frequency.

C. Case Studies

Now that we have examined the methods for production and selection of somatic
hybrids, let us look at some examples of somatic hybridization from the literature. A
summary of somatic hybrids which have yielded plants is presented in Table 1.

1. Arabidobrassica

One of the best known examples of somatic hybridization is *Arabidobrassica*.[16,17]
This is an intergeneric hybrid produced from fusing protoplasts of *Arabidopsis thaliana* and *Brassica campestris,* both members of the Cruciferae family. Plants produced
from such fusions contained genetic material from both parents but were, in most
cases, what is termed asymmetric hybrids. That is, the regenerates resembled one parent more than another and, likewise, contained only a small portion of the chromosomal complement from one parent. *B. campestris* cells contained 20 chromosomes
and the octaploid callus line of *A. thaliana* used here had 40 chromosomes. The resulting hybrids contained anywhere from 35 to 80 chromosomes, usually heavily favoring
one parent over another, the remainder presumably lost by chromosomal elimination.

2. Tomato-Potato Hybrids

Another well known example of somatic hybridization comes from the work of
Melchers et al.[21] and Shepard et al.[22] Tomato (*Lycopersicon esculentum*) and potato
(*Solanum tuberosum*) are both members of the Solanaceae family but remain sexually
incompatible. Melchers et al.[21] were the first to demonstrate that somatic hybrids of
tomato and potato are asymmetric, as in the case of *Arabidobrassica,* and hybrids have
been dubbed "topato" or "pomato" depending on which parent the hybrid resembles.
Although the hybrids were asymmetric, they did display morphological characteristics
of both parents and an electrophoretic analysis of the large subunit of ribulose-1,5-
bisphosphate carboxylase (RuBPCase) indicated that some contained the chloroplast
genome of tomato whereas others contained the chloroplast genome of the potato.
These hybrids all showed chromosomal elimination to some degree and, although some
tuber-like stolons were formed, there were no fertile flowers or fruit set.

Likewise, Shepard et al.[22] was able to show the formation of asymmetric hybrids from fusion of tomato and potato protoplasts. Somatic hybrids demonstrated the growth habit of a potato-like vine although leaf serrations and lobing were intermediate between the two parents. The hybrids grew vigorously and produced large amounts of anthocyanins in the stem and the underside of leaves, reminiscent of tomato. Small to medium-sized white tubers were formed and, on some hybrids, sterile yellow fruit were set which had a distinct "tomato-like odor". In addition, these hybrids were found to contain the small subunit of RuPBCase from both potato and tomato which could be distinguished by isoelectric focusing.

III. FUTURE PROSPECTS

In addition to the two examples presented before, there are many other somatic hybrids which have been produced (see Table 1). What are the lessons to be learned from these examples? Several trends have become clear from this work. First, intergeneric somatic hybrids can be produced between species which are sexually incompatible. In many cases, whole plants can be regenerated which are intermediate, in many respects (e.g., morphology, growth characteristics, biochemistry), between the two parental types. Secondly, in all of the cases of intergeneric, and in some of the cases of interspecific, somatic hybridization, the plants produced were infertile. Although some, like the topatoes, were capable of setting fruit, others were incapable of forming any flowers whatsoever. And thirdly, in most cases, intergeneric somatic hybrids would yield whole plants only if the parents were from the Solanaceae family. This has led to the generalization that only plants from this family are capable of regeneration. However, recent advances in the regeneration of non-Solanaceous plants from protoplasts effectively dispels this notion.[5]

So what are the future prospects of somatic hybridization as an alternative to traditional plant breeding? We believe that the future looks quite promising, with some limitations. The first such limitation concerns the phylogenetic distance of the parents to be fused. It is clear that if one is interested in obtaining fertile offspring, then the parents must be closely related, preferably within the same genus. Secondly, one must work with a system that is capable of regeneration of whole plants from protoplasts. Although this list is limited, it is growing very rapidly and, at present, there are a dozen agriculturally important crops that have been regenerated. Notably absent from this list, however, are the cereals. And thirdly, it is important to define and identify precisely those traits or characteristics one wants to combine in somatic hybrids. Tremendous variation exists in gene pools and several recent examples have exploited this fact to yield novel somatic hybrids. For example, Evans et al.[26] and Uchimiya[27] have demonstrated enhanced resistance, especially to infection by TMV, in somatic hybrids of *Nicotiana tabacum* + *N. nesophila* and *N. tabacum* + *N. glutinosa*. Likewise, Butenko et al.[28] were able to produce potato somatic hybrids between *Solanum tuberosum* + *S. chacoense,* that were resistant to potato × virus.

This last example, namely, the hybrid between *S. tuberosum* + *S. chacoense,* illustrates an important point. That is, that crosses between commercially important species (e.g., *S. tuberosum*) and wild ancestral types (e.g., *S. chacoense*) offers many advantages.[2] In general, ancestral plants often have envious qualities like resistance to disease or improved protein content which is inadvertently lost during modern domestication. For example, an ancestor of wheat contains 30% protein on a dry weight basis, more than twice that found in modern varieties.[2] As pointed out by Shepard et al.[22] there is still a great many opportunities to increase the disease resistance in plants like potato. One of the more primitive species of potato (*S. etuberosa*), for example, is resistant to the potato leaf roll virus[29] and the potential exists to pass on this trait via somatic hybridization since these two species are sexually incompatible. Thus, such an approach might allow for the rather rapid introduction of resistance genes into current germplasm pools.

It is clear that, with the rapid progress in the area of plant molecular biology, one will be able to prescribe, in the future, a precise "gene therapy" to correct particular deficiencies in plants.[30] This is especially true in light of the increasing number of plant genes identified and cloned and with improved vector systems for plant transformation. In the immediate future, however, we should expect to see an increasing dependence on somatic hybridization as a means of producing novel genotypes between sexually incompatible parents.

IV. SUMMARY

In the near future, somatic hybridization may be used to create novel hybrid plants which express beneficial traits of two or more sexually incompatible parents. By fusing closely related species, one can avoid the problems associated with the fusion of phylogenetically distant species and insure that hybrid offspring will be viable. Such somatic hybrids have the potential of combining such valuable traits as resistance to disease, ability to detoxify herbicides, tolerance to heat and frost, and enhanced resistance to salt and drought stress, thereby increasing both the environmental range as well as the yield potential of agriculturally important plants.

REFERENCES

1. Borlaug, N. E., Contributions of conventional plant breeding to food production, *Science*, 219, 689, 1983.
2. Tudge, C., The future of crops, *New Sci.*, 98, 547, 1983.
3. Sprague, J. J., Combining genomes by conventional means, in *Plant Improvement and Somatic Cell Genetics*, Vasil, I. K., Scowcroft, W. R., and Frey, K. J., Eds., Academic Press, New York, 1982, chap. 5.
4. Flores, H. E., Sawhney, R. K., and Galston, A. W., Protoplasts as vehicles for plant propagation and improvement, in *Advances in Cell Culture*, Vol. 1, Maramorosch, K., Ed., Academic Press, New York, 1981, 241.
5. Evans, D. A., Agricultural applications of plant protoplast fusion, *Biotechnology*, 1, 253, 1983.
6. Senda, M., Takeda, J., Abe, S., and Nakamura, T., Induction of cell fusion of plant protoplasts by electrical stimulation, *Plant Cell Physiol.*, 20, 1441, 1979.
7. Zimmermann, U., Vienken, J., and Pilwat, G., Development of drug-carrier systems: electric field induced effects in cell membranes, *Bioelectrochem. Bioenerg.*, 7, 553, 1980.
8. Zimmermann, U. and Scheurich, P., Fusion of *Avena sativa* mesophyll protoplasts by electrical breakdown, *Biochim. Biophys. Acta*, 641, 160, 1981.
9. Zimmermann, U. and Scheurich, P., High frequency fusion of plant protoplasts by electric fields, *Planta*, 151, 26, 1981.
10. Gaynor, J. J., Electrofusion of plant protoplasts, in *Handbook of Plant Cell Culture*, Vol. 4, Evans, D. A., Sharp, W. R., and Bravo, J. E., Eds., Macmillan, New York, 1985.
11. Zimmermann, U., Electric field-mediated fusion and related electrical phenomena, *Biochim. Biophys. Acta*, 694, 227, 1982.
12. Zimmermann, U. and Vienken, J., Electric field induced cell-to-cell fusion, *J. Memb. Biol.*, 67, 158, 1982.
13. Bates, G. W., Gaynor, J. J., and Shekhawat, N. S., Fusion of plant protoplasts by electric fields, *Plant Physiol.*, 72, 1110, 1983.
14. Harms, C. T. and Potrykus, I., Fractionation of plant protoplast types by iso-osmotic density gradient centrifugation, *Theor. Appl. Genet.*, 53, 57, 1978.

15. Flores, H. E., Thier, A., and Galston, A. W., In vitro culture of grain and vegetable amaranths (*Amaranthus* spp.), *Am. J. Bot.,* 69, 1049, 1982.
16. Gleba, Y. Y. and Hoffmann, F., "Arabidobrassica": plant-genome engineering by protoplast fusion, *Naturwissenschaften,* 66, 547, 1979.
17. Hoffmann, F. and Adachi, T., "Arabidobrassica": chromosomal recombination and morphogenesis in asymmetric intergeneric hybrid cells, *Planta,* 153, 586, 1981.
18. Krumbiegel, G. and Schieder, O., Selection of somatic hybrids after fusion of protoplasts from *Datura innoxia* Mill. and *Atropa belladonna* L., *Planta,* 145, 371, 1979.
19. Dudits, D., Hadlaczky, G. Y., Bajszar, G. Y., Koncz, C. S., and Lazar, G., Plant regeneration from intergeneric cell hybrids, *Plant Sci. Lett.,* 15, 101, 1979.
20. Dudits, D., Hadlaczky, G., Lazar, G., and Haydu, Z., Increase in genetic variability through somatic cell hybridization of distantly related plant species, in *Plant Cell Cultures: Results and Perspectives,* Sala, F., Parisi, B., Cella, R., and Ciferri, O., Eds., Elsevier/North-Holland, Amsterdam, 1980, 207.
21. Melchers, G., Sacristan, M. D., and Holder, S. A., Somatic hybrid plants of potato and tomato regenerated from fused protoplasts, *Carlsberg Res. Commun.,* 43, 203, 1978.
22. Shepard, J. F., Bidney, D., Barsby, T., and Kemble, R., Genetic transfer in plants through interspecific protoplast fusion, *Science,* 219, 683, 1983.
23. Li, X. H., Li, W. B., and Huang, M. J., Somatic hybrid plants from intergeneric fusion between tobacco tumor B6S3 and *Petunia hybrida* W43 and expression of LpDH, *Sci. Sin.,* 25, 611, 1982.
24. Gleba, Y. Y. and Evans, D. A., Genetic analysis of somatic hybrid plants, in *Handbook of Plant Cell Culture,* Evans, D. A., Sharp, W. R., Ammirato, P. V., and Yamada, Y., Eds., Macmillan, New York, 1, 322, 1983.
25. Nagao, T., Somatic hybridization by fusion of protoplasts. III. Somatic hybrids of sexually incompatible combinations *Nicotiana tabacum* + *Nicotiana repanda* and *Nicotiana tabacum* + *Salpiglossis sinuata,* *Jpn. J. Crop Sci.,* 51, 35, 1982.
26. Evans, D. A., Flick, C. E., Kut, S. A., and Reed, S. M., Comparison of *Nicotiana tabacum* and *Nicotiana nesophila* hybrids produced by ovule culture and protoplast fusion, *Theor. App. Genet.,* 63, 193, 1982.
27. Uchimiya, H., Somatic hybridization between male sterile *Nicotiana tabacum* and *Nicotiana glutinosa* through protoplast fusion, *Theor. Appl. Genet.,* 61, 69, 1982.
28. Butenko, R., Kuchko, A., and Komarnitsky, I., Some features of somatic hybrids between *Solanum tuberosum* and *Solanum chacoense* and its F₁ sexual progeny, in *Plant Tissue Culture 1982, Proc. 5th Int. Congr. Plant Tissue and Cell Culture,* Fujiwara, A., Ed., Japanese Association of Plant Tissue Culture, Tokyo, Japan, 1982, 643.
29. Smith, N., New genes from wild potatoes, *New Sci.,* 98, 558, 1983.
30. Barton, K. A. and Brill, W. J., Prospects in plant genetic engineering, *Science,* 219, 671, 1983.
31. Gaynor, J. J. and Bates, G. W., Fusion of plant protoplasts with an electric field, *J. Cell Biol.,* 95, 110a, 1982.

Chapter 3

USE OF PLANT TISSUE CULTURE TECHNIQUES IN PLANT BREEDING

Chee-Kok Chin

TABLE OF CONTENTS

I. INTRODUCTION

Advances in plant breeding undoubtedly are one of the major factors contributing to the success in agriculture in recent history. Through breeding, crops that are high yielding, stress resistant, pest resistant, and possess other good qualities are produced. The standard method used in plant breeding has been based on genetic recombination through sexual crosses. Several steps are used in the conventional breeding work. These are (1) collection of germplasm, (2) selection of plants with desired characteristics, (3) crossing the plants with desired characteristics, and (4) selection of hybrids or pure lines with combined desired characteristics.

Although plant breeding has produced many improved crop plants, there are limitations to this approach. The main limitations are the availability of germplasm and the incompatibility of certain crosses. Thus, it is impossible to introduce certain traits into crop plants if no plants with genes for such traits are available, or if plants with these genes are not sexually compatible with the crop plants of interest.

Since practically all plants have some degree of heterozygosity, the formation of gametes from meiosis segregation results in a very heterogeneous population of gametes. Because a breeder has no control over the segregation of gametes and thus the subsequent recombination of the cross, the breeder has to rely on many and often repeated crossings to obtain a plant with a certain combination of traits.

To overcome some of these obstacles, many approaches have been attempted, e.g., to produce new genes by promoting mutation, to bypass incompatibility by manipulation of environment and flower structure, etc. So far, the success of this manipulation has been limited. In recent years, the use of plant tissue culture techniques to assist plant breeding has been explored and the results appear to be promising. The following is a survey of some of the possible uses of plant tissue culture techniques in plant breeding.

II. HAPLOID PLANTS

One way that plant breeders reduce or eliminate the degree of heterogeneity in gametes is to use homozygous lines in crossing. The conventional approach in developing homozygous lines is through controlled self-pollination. This is a time-consuming procedure, especially for plants with a long generation time, since many generations of selfing are needed to ensure an adequate level of homozygosity. In addition, some naturally cross-pollinated species have difficulty in selfing due to self-incompatibility.[1]

Another approach in developing homozygous lines is to utilize haploid plants. These plants could be converted to homozygous diploids by chromosome doubling, a process that may be induced by treatment with chemicals such as colchicine.[2] One problem with using haploid plants is that these plants rarely exist in nature. Many methods have been employed to produce them. These include interspecific hybridization,[3] crossing of plants with imbalanced chromosome makeup,[4] use of irradiated pollen,[5] drastic temperature changes,[6] and delayed pollination.[6] Other than a few exceptions, none of these methods has been found to be sufficiently and consistently effective.

Unlike haploid plants, haploid cells present as gametes are common in nature. The gametes of many plants can be cultured on nutrient medium and haploid plants regenerated from them.[7,8] Pollen grains, the male gametophytes, are usually used for this purpose because they are produced in large quantities and are easily obtainable. Pollen grains normally develop into pollen tubes and sperms. In pollen culture, the development is manipulated to switch from pollen germination and sperm production to cell proliferation and regeneration.

Many factors, e.g., medium composition, cultural conditions, and physiological state of the donor plants are known to influence the success of pollen culture. In some instances, the success can be aided by allowing the pollen grains to remain undisturbed within the anther. This method, called anther culture, is technically simpler. However, there is one possible disadvantage of this method, i.e., plants may regenerate not only from pollen but also from the somatic cells of the anther producing both haploid and diploid plants. Consequently, screening is needed to root out the plants regenerated from somatic cells. For this reason, production of haploid plants by culture of isolated pollen grains are preferred.

To obtain homozygous diploid plants, the haploid plants from pollen or anther culture are treated with colchicine. Alternatively, the anther may be treated with colchicine at the time of the first pollen mitosis prior to culture. The diploidized cells then will give rise to homozygous plants directly.

Using pollen and anther culture, haploid and homozygous plants of many crop species have been produced.[8] However, success of many others, including many important crop species, has yet to be demonstrated. It is known that some genotypes are easier to culture than others. The reason for this, however, is not known.

III. RECOVERY OF MUTANTS

Since conventional plant breeding involves recombining the genes, the availability of genetic diversity or germplasm is essential. Genetic variation has its origin in mutation. Mutation is a random process and may happen to any cell of a plant resulting in production of a chimera, i.e., a plant with cells of more than one genetic makeup. Only mutation in reproductive cells and not in somatic cells will transmit the genetic change to the next generation. Many somatic mutations, therefore, are lost in nature unless the plants are propagated vegetatively. This, together with the fact that mutation occurs only at a very low frequency, limits the availability of genetic diversity. One way to recover the somatic mutants is to culture these cells and regenerate them into plants. However, mutant cells are not often visually distinguishable from the normal cells. Recovery of them, therefore, depends on regenerating all cells into plants and then examining their traits. This will involve regenerating and growing a large number of plants, many of which may not be mutants. This time-consuming process can be avoided if screening procedures are available to select for the mutants.

It appears that certain mutations can be identified by rather simple screening methods. For example, resistance to herbicides, pesticides, pathogen toxins, and salts may be screened for by incorporation of the substance of interest in the medium. The presence of the substance would allow the growth of the resistant cells but not the susceptible ones. Using this method, mutants resistant to different diseases have been reported. For example, clones of tobacco resistant to wildfire disease caused by *Pseudomonas tobaci*,[9] sugar cane resistant to eyespot disease caused by *Helminthosporium sacchari*,[10] and potato resistant to late blight disease caused by *Phytophthora infestans*[11] have been selected by this method.

For higher plants, the expression of traits normally changes with plant development stages. Thus, one potential pitfall in using cultured cells in screening is that traits expressed by cultured cells may not be expressed by the whole plant. For this reason, it is essential that the validity of the screening methods be confirmed before they are used in actual screening. Because the expression of many traits changes with development, certain traits do not lend themselves to screening at the cellular stage. This is true for many important agronomical traits such as plant size, leaf canopy characteristics, flowering behavior, yield, etc., that are expressed only by the whole plant. Nevertheless, it is possible that certain correlation may exist between some important whole plant traits and cellular characteristics. If such correlations exist and can be identified, they may be used in screening for those whole plant traits.

Many chemicals, e.g., ethyl methanesulfonate (EMS), ethyl ethanesulfonate (EES), bromouracil, nitrous acid, etc., and certain irradiations, e.g., ultraviolet, X-ray, and γ-ray, are known to induce mutation. Since these factors can be readily applied to cells in culture, they can be used to increase the frequency of mutation of the cultured cells.

Because mutation is recessive in nature and somatic cells of most crop plants are either diploid or polyploid, the mutation will be expressed only if all alleles are affected. The probability of this, of course, is very low. Thus, it can be seen that even with cell culture techniques many mutations will not be detected. A possible solution to this problem is to use haploid cell culture initiated from the pollen and anther. With haploid cells there is only one set of chromosomes present. As a result, there will be no masking of the mutated allele by another allele. The mutant haploid cells can be diploidized later with chemical treatment and then regenerated into homozygous diploid mutant plants.

IV. IN VITRO FERTILIZATION

Breeding through conventional genetic recombination is often hampered by incompatibility which is encountered in certain selfing and most crosses among genotypes that are not closely related, such as interspecific and intergeneric crosses. Different kinds of incompatibility have been observed; for example, failure in pollen germination, in pollen tube growth, and in pollen tube penetration into the ovule. It is believed that the presence of proteins or glycoproteins on the incompatible pollen may elicit reactions in the stigma, style, or ovule, which subsequently leads to the failure in pollen germination and arrest in pollen tube growth or penetration. The problems associated with the stigma and style tissues sometimes can be bypassed by direct injection of a pollen slurry into the ovary[16] to bypass incompatibility. Another approach is in vitro fertilization. In this method, an ovary intact on the placenta with the style removed is cultured and the ovule mass is then fertilized with pollen grains.[17] Removal of the stigma and style eliminates the incompatibility encountered in these tissues. This method has been used to produce fertilization in cases that would be otherwise incompatible.

V. EMBRYO CULTURE

Sexual incompatibility is not the only limiting factor of many crosses. Occasionally, although fertilization is successful, the resulted embryo fails to develop to maturity and aborts. This is most often found in intergeneric crosses such as barley × rye[12] and interspecific crosses such as *Brassica oleracea* L. × *B. campestris* L.[13] Embryo abortion is also found in certain early ripening fruits such as peaches and apricots. In many cases, the collapse of the embryo is associated with degeneration of the endosperm. Endosperm is known for its function as a source of nutrition for the embryo. It is believed that collapse of the endosperm deprives the embryo of nutrients and hence leads to abortion. This view is supported by the demonstration that many of these embryos, when excised and placed in culture, will continue to grow to produce viable seedlings.[14] Regardless of what actually is the cause of the abortion, many of these embryos can be rescued from abortion by culturing them in vitro. Indeed, embryo cultures are being used in many breeding programs for this purpose.

Not all the aborted embryos can be successfully rescued by embryo culture. An important problem may be that the medium used in the culture is not appropriate and some of the essential nutrients are missing from it. A simple solution to this problem is to place the embryo on immature endosperm excised from the developing seed of one of the parents. The endosperm serves as a nurse tissue for the hybrid. With this method, the survival of the embryos from many interspecific crosses, e.g., Triticale crosses, has been found to improve significantly.[15]

VI. PROTOPLAST FUSION

In the cases where in vitro placenta fertilization is ineffective, the incompatibility may be bypassed by protoplast fusion. Protoplasts are used because the cell wall is a barrier to fusion. The method involves the digestion of cell wall with a mixture of cell wall degrading enzymes, viz., cellulases, hemicellulases, and pectinases. Fusion of protoplasts from different sources does not occur spontaneously but has to be induced. Several factors, including calcium ions at high pH,[18] sodium nitrate,[19] polyethylene glycol (PEG),[20,21] and electric current,[22] have been reported to induce fusion. Results obtained with PEG have been most consistent and this approach is most widely used currently. The PEG induces the protoplasts to agglutinate and, when the PEG is later diluted, the protoplasts fuse. Fusion induced by PEG is nonspecific. It allows protoplasts from different sources to fuse. Recently, a technique utilizing short pulse of direct current to induce fusion has been developed.[22] This method, which is termed electrofusion, is very efficient and apparently is more gentle to the protoplasts than the PEG treatment. The indications are that this will become a very important cell fusion method (see Chapter 2).

Fusion and subsequent cell division have been reported for closely and distantly related plants.[23] Thus, close relatedness is not essential for induced fusion. Fusion, however, does not always lead to production of amphiploid cells, i.e., cells with chromosomes from both parents.[24] The problem is that sometimes fusion is followed by the loss of chromosomes[25] in subsequent cell divisions. The number of chromosomes that are eliminated varies. It may range from one to, in some extreme cases, the whole set of chromosomes of one parent.[26] In general, the degree of elimination increases as the relatedness of the parents decreases. Apparently, the parental genomes are not the only factors affecting the chromosome loss since fusion of the same two parents may lead to heterokaryocytes of a range of chromosome numbers.[24]

The complete loss of chromosomes of one parent does not necessarily mean the complete loss of the genetic information from that parent. For example, peroxidase isozyme patterns of both parents were retained in hybrids of *Petunia hybrida* and *Parthenocissus*, even though the hybrid completely lost the chromosome of the latter.[26] Apparently, some genes or chromosome segments of one parent must have been incorporated into the other. It appears then, for the purpose of introducing certain genes from one plant to another, as long as the plant genome of the latter and the genes of interest from the former are maintained, the purpose will be served. This will not be the case, of course, if the genomes or the genes of interest are eliminated. Thus, whether the chromosome loss is a problem depends on what chromosomes or genes are lost.

Somatic fusion not only results in combination of nuclei but also cellular organelles. Since organelles like mitochondria and chloroplasts carry genetic information, the mixing of them would affect the genetic makeup of the hybrids. Hybrids of only nonnuclear genes can also be produced by a modified protoplast fusion method. For example, Raveh et al.[27] used a method of fusing X-irradiated protoplasts with nonirradiated ones. X-irradiation inactivates the nuclei. The fusion thus results in a hybrid (cybrid) with nucleus from the nonirradiated parent but cytoplasm from both parents. Already, a number of agriculturally important genes, e.g., resistance to herbicide, and male sterility are known to be carried by the organelles.[28,29] Thus, transfer of chloroplasts and mitochondria may have important applications in plant improvement.

One potential handicap to the use of cytoplasmic fusion is that it is commonly observed that, following fusion, there is a sorting out of chloroplasts resulting in eliminating those from one of the two parents.[30,31] Nevertheless, there are exceptions.[32] The usefulness of cytoplasmic fusion will then depend on whether there is sorting out of the mixed organelles.

VII. TRANSFORMATION

A common approach in plant breeding is to start with a plant with many good traits and then introduce whatever desirable traits this plant is lacking from another plant by crossing. This task is complicated by the fact that the plant that carries those desirable traits often also carries some undesirable traits which will be incorporated into the progenies. Time- and labor-consuming backcrossing and further crossing are needed to rid the hybrids of the undesirable traits but retain the desirable ones. Normally, the severity of this problem increases as the relatedness of parent plants decreases. Thus, incompatibility is not the only problem in crossing distantly related plants. For example, using protoplast fusion, it has been possible to cross the otherwise sexually incompatible potato and tomato;[24] however, the hybrids acquire a mixture of characteristics from both parents, and consequently lack the crop values of either parents.

The problem of incorporating unwanted genes into the hybrids can be avoided if the desirable genes and not the whole genome can be selectively introduced by transformation. To achieve this, a combination of recombinant DNA and tissue culture techniques is needed. With the recombinant DNA technique, a particular gene is isolated and incorporated into a plasmid such as the Ti plasmid from *Agrobacterium tumefaciens* to produce a hybrid plasmid. Agrobacteria with the Ti plasmids have been transforming plant cells in nature resulting in "crown gall" tumor. The hybrid plasmid is cloned within a bacterial cell. The plasmids are then isolated, purified, and introduced into plant protoplasts which are later regenerated into plants. The advantage of this approach is that one may engineer a plasmid to carry a particular gene or number of genes. Thus, introgression is achieved with a successful transformation. In addition, once a plasmid carrying a gene of interest such as a disease resistant gene is synthesized, it can be used to transform different plant protoplasts and both the sexual incompatibility and repeated backcrossing for introgression can be avoided.

At present, identification and cloning of plant genes and transformation of plant cells are only at the infant stage. Many difficult problems need to be solved before the potential of this approach can be realized.

VIII. CONCLUDING REMARKS

There is no doubt that tissue culture techniques can help to overcome many problems faced by conventional plant breeding. However, so far, only some of these techniques such as embryo culture and somaclonal variants selection are being used and then only in limited ways. One reason that these techniques have not been more fully exploited is probably because with many of these techniques there are still technical difficulties to be resolved. For example, although many plants can be readily regenerated from callus or cell through organogenesis or embryogenesis, many others fail to do so. The latter category includes economically important crops, such as soybean and many cereal species.[33] The genotypic differences in response are not only restricted to the regeneration of plants but are observed with all the techniques mentioned above. Also, even with the same genotype, the responses may vary with the sources of tissues, e.g., stem or leaf; the developmental stages of the donor plants, e.g., juvenile or mature; physiological status of the donor plants, e.g., whether or not they are vigorously growing, etc. Sometimes the causes for the differences are more obscure. For example, regeneration of certain wheat species, even though it can be done, is only at very low frequencies and is very inconsistent.[34] In these cases, where cultures come from the same source and are cultured under the same conditions, the reason why some regenerate and others do not is not known.

Generally, a culture technique is developed through a trial and error approach usually involving the examination of various media compositions, especially in terms of the types and concentrations of growth regulators, methods of culture, cultural conditions, tissue sources, etc. This approach works well for many plants but not for many others.

At present, little is known about the biochemical and physiological basis for the differences in response. Since differences exist even among cells of a single plant, it is reasonable to assume that they are caused by cytoplasmic factors. The question remains why some cells are able to express a certain developmental pattern such as regeneration whereas others do not. Could it be that those which can do so possess certain substances to activate the expression and those which cannot do not? If the cytoplasm environment is a determining factor, the problem may be corrected by modifying the culture medium which in turn may alter the cytoplasmic environment. Indeed, it is known that, in general, auxin promotes root formation and cytokinin shoot formation. Regeneration of plants is usually obtained by a proper combination of auxin and cytokinin.[35] However, some plants cannot be regenerated by various combinations of auxin and cytokinin. Does this mean that other promoting substances are needed? Little is known about other substances capable of promoting plant regeneration except that it is commonly observed that tissues with veins regenerate plants more readily than those without, and plants are most frequently regenerated at the cut ends of the veins where some endogenous substances may accumulate.

Alternatively, those tissues that fail to regenerate may possess certain substances inhibitory to regeneration. Substances inhibitory to regeneration have been reported, e.g., embryogenesis is inhibited by auxin or gibberellin. Thus, removal of auxin is often used to induce embryogenesis.[33] Gibberellin is also known to inhibit root and shoot formation. The growth retardant, ancymidol, which blocks the synthesis of gibberellin, is found to effectively promote shoot and root formation of asparagus.[36] Also, increases in plant regeneration by charcoal has been observed. The effect of charcoal has been attributed to its ability to remove inhibitory substances produced by the cultures.[37]

Along with growth substances, mineral salts in the medium have attracted much attention. In fact, a medium is generally known for its salt composition. The composition and concentration of salts in different media may vary widely.[38] Although a medium is usually developed for the culture of a particular type of tissue, it is not uncommon to find that it can be used to culture a range of tissues or cells.[33] Also, some cultures, e.g., tobacco callus, can grow satisfactorily in several different media. This leads to a rather common practice that a medium is arbitrarily selected for the culture of a tissue. This may not be a good practice and the practice may be responsible for many culture failures because it is known that there are tissues that grow well only in certain media but not in others.[39] At present, the physiological relationship between medium and tissue is poorly understood. Also, it is not known why certain tissues are rather indiscriminate whereas others are selective in their requirements of mineral salts.

Minerals affect not only growth but also development in whole plants. For example, nitrogen generally encourages production of foliage and phosphorus encourages roots. In cell culture, nitrogen salts KNO_3 or NH_4NO_3 are observed to stimulate embryogenesis.[40] Studies are required to find out more about the effect of mineral salts on development in vitro. It is possible that some developmental problems that, so far, are not solvable by the manipulation of growth regulators can be solved by a combination of manipulation of mineral salts and growth regulators.

One component of the culture medium the influence of which may be underestimated is the carbon source. Sucrose, the most abundant sugar found in most plants, is used in practically all culture media. For many tissues, sucrose supports much better growth than other sugars including glucose and fructose.[41,42] It appears that the differences in the growth-supporting abilities of different sugars are related to their absorption[43] and metabolism.[44] The superiority of sucrose, however, is not universal. Generally, glucose is significantly superior to sucrose in supporting the growth of monocotyledonous roots.[44] It is possible that the selection of the wrong carbon source for the medium may result in failure in culture.

It is believed that, in general, somatic plant cells are genetically totipotent and have the potential to exhibit a whole series of growth and developmental processes. Since the expression of these processes is dependent on the cytoplasmic and ambient environment, it is important to find out not only the ambient conditions but also the cytoplasmic milieus for various growth and developmental processes. With such information, maybe we will know more about how to culture and manipulate different plant cells, tissues, and organs for breeding purposes. Without such information, as we essentially are today, the use of many of the techniques mentioned above for practical plant improvement are handicapped.

REFERENCES

1. Lewis, D., Incompatibility in flowering plants, *Biol. Rev.*, 24, 472, 1949.
2. Jensen, C. J., Chromosome doubling techniques in haploids, in *Haploid in Higher Plants, Advances and Potential,* Kosha, K. J., Ed., University of Guelph Press, Ontario, Canada, 1974, 153.
3. Kao, K. N. and Kasha, K. J., Haploidy from interspecific crosses with tetraploid barley, in *Barley Genetics*, Vol. 2, Nilan, R. A., Ed., Washington State University Press, Pullman, Wash., 1969, 82.
4. Emmerling, M. H., A comparison of x-ray and ultraviolet effects on chromosomes of *Zea mays, Genetics*, 40, 697, 1955.
5. Stadler, L. J., Notes on haploids, *Maize Genet. Coop. Newsl.,* 14, 27, 1940.
6. Chase, S. S., Monoploids and monoploid derivatives of maize (*Zea mays* L.), *Bot. Rev.*, 35, 117, 1969.
7. Juha, S. and Maheshwari, S. C., *In vitro* production of embryos from anthers of *Datura, Nature (London),* 204, 497, 1964.
8. Chu, C., Haploids in plant improvement, in *Plant Improvement and Somatic Genetics,* Vasil, I. K., Scowcroft, W. R., and Frey, K. J., Eds., Academic Press, New York, 1982, 19.
9. Carlson, P. S., Methionine sulfoximine-resistant mutants of tobacco, *Science*, 180, 1366, 1973.
10. Heinz, D. J., Krishnamurthi, M., Nickell, L. G., and Maretzki, Cell, tissue and organ culture in sugarcane improvement, in *Applied and Fundamental Aspects of Plant Cell, Tissue and Organ Culture,* Reinert, J. and Bajaj, Y. P. S., Eds., Springer-Verlag, New York, 1977, 3.
11. Shepard, J. F., Bidney, D., and Shahin, E., Potato protoplasts in crop improvement, *Science*, 208, 17, 1980.
12. Brink, R. A., Copper, D. C., and Ausherman, L. E., A hybrid between *Hordeum jubatum* and *Secale cereale, J. Heredity,* 35, 67, 1944.
13. Nishi, S., Kawata, J., and Toda, M., On the breeding of interspecific hybrids between two genomes, 'c' and 'a' of *Brassica* through the application of embryo culture techniques, *Jpn. J. Breed,* 8, 215, 1959.
14. Raghaven, V., Applied aspects of embryo culture, in *Applied and Fundamental Aspects of Plant Cell, Tissue, and Organ Culture,* Reinert, J. and Bajaj, Y. P. S., Eds., Springer-Verlag, Berlin, 1977, 412.
15. Williams, E., Verry, I. M., and Williams, W. M., Use of embryo culture in interspecific crosses, in *Plant Improvement and Somatic Genetics,* Vasil, I. K., Scowcroft, W. R., and Frey, K. J., Eds., Academic Press, New York, 1982, 119.

16. Kanta, K. and Maheshwari, P., Intraovarian pollination in some *Papaveraceae, Phytomorphology,* 13, 215, 1964.
17. Rangaswamy, N. S. and Shivanna, K. P., Induction of gametic compatibility and seed formation in axenic cultures of a diploid self-incompatible species of petunia, *Nature (London),* 16, 937, 1967.
18. Keller, W. A. and Meichers, G., The effect of high pH and calcium on tobacco leaf protoplast, *Z. Naturforsch.,* 28C, 737, 1973.
19. Power, J. B., Cummins, S. E., and Cocking, E. C., Fusion of isolated protoplasts, *Nature (London),* 225, 1016, 1970.
20. Kao, K. N. and Michayluk, M. R., A method for high frequency intergeneric fusion of plant protoplasts, *Planta,* 115, 355, 1974.
21. Wallin, A., Glimelius, K., and Eriksson, T., The induction of aggregation and fusion of *Daucus glycol. Z. Pflanzenphysiol.,* 74, 64, 1974.
22. Zimmermann, U. and Sheurich, P., High frequency fusion of plant protoplasts by electric fields, *Planta,* 151, 26, 1981.
23. Schieder, O. and Vasil, I. K., Protoplast fusion and somatic hybridization, in *Perspectives in Plant Cell and Tissue Culture,* Vasil, I. K., Ed., Academic Press, New York, 1980, 137.
24. Shepard, J. F., Bidney, D., Barsby, T., and Kemble, R., Genetic transfer in plants through interspecific protoplast fusion, *Science,* 219, 683, 1983.
25. Kao, K. N., Chromosomal behavior in somatic hybrids of soybean — *Nicotiana glauca, Mol. Gen. Genet.,* 150, 225, 1977.
26. Power, J. B., Freason, E. M., Haywood, C., and Cocking, E. C., Some consequences of the fusion and selective culture of petunia and *Parthenocissus* protoplasts, *Plant Science Lett.,* 5, 197, 1975.
27. Reveh, D., Huberman, E., and Galun, E., *In vitro* culture of protoplasts: use of feeder techniques to support division of cells placed at low densities, *In Vitro,* 9, 216, 1973.
28. Pfister, K., Steinback, K. E., Gardner, G., and Arntzen, C. J., Photoaffinity labeling of an herbicide receptor protein in chloroplast membranes, *Proc. Natl. Acad. Sci. U.S.A.,* 78, 981, 1981.
29. Belliard, G., Vedel, F., and Pelletier, G., Mitochondrial recombination in cytoplasmic hybrids of *Nicotiana tobacum* by protoplast fusion, *Nature (London),* 281, 401, 1979.
30. Kung, S. D., Gray, J. C., Wildman, S. G., and Carlson, P. S., Polypeptide composition of fraction I protein from parasexual hybrid plants in the genus *Nicotiana, Science,* 187, 353, 1975.
31. Chen, K., Wildman, S. G., and Smith, H. H., Chloroplast DNA distribution in parasexual hybrids as shown by polypeptide composition of fraction I protein, *Proc. Natl. Acad. Sci. U.S.A.,* 74, 5109, 1977.
32. Galun, E., Somatic cell fusion for inducing cytoplasmic exchange: a new biological system for cytoplasmic genetics in higher plants, in *Plant Improvement and Somatic Cell Genetics,* Vasil, I. K., Scowcroft, W. R., and Frey, K. J., Eds., Academic Press, New York, 1982, 205.
33. Evans, D. A., Sharp, W. R., and Flick, C. E., Plant regeneration from cell cultures, *Hortic. Rev.,* 3, 214, 1981.
34. Bhojwani, S. S. and Hayward, C., Some observations and comments on tissue culture of wheat, *Z. Pflanzenphysiol.,* 85, 341, 1977.
35. Miller, C. O. and Skoog, F., Chemical control of bud formation in tobacco stem segments, *Am. J. Bot.,* 40, 768, 1953.
36. Chin, C., Promotion of shoot and root formation in asparagus *in vitro* by ancymidol, *HortScience,* 17, 590, 1982.
37. Bajaj, Y. P. S., Reinert, J., and Heberle, E., Factors enhancing *in vitro* production of haploid plants in anthers and isolated microspores, in *A la Memoire de Georges Morel,* Gautheret, R. J., Ed., Masson and Cie, Paris, 1976.
38. Gamborg, O. L., Murashige, T., Thorpe, T. A., and Vasil, I. K., Plant tissue culture media, *In Vitro,* 12, 473, 1976.
39. Wolfe, D. E., Eck, P., and Chin, C., Evaluation of seven media for micropropagation of highbush blueberry, *HortScience,* 18, 703, 1983.
40. Tazawa, M. and Reinert, J., Extracellular and intracellular chemical environments in relation to embryogenesis *in vitro, Protoplasma,* 68, 157, 1969.
41. White, P. R., Potentially unlimited growth of excised tomato roots, *Plant Physiol.,* 9, 585, 1934.
42. Ferguson, J. D., Street, H. E., and David, S. B., The carbohydrate nutrition of tomato roots. V. The promotion and inhibition of excised root growth by various sugars and sugar alcohols, *Ann. Bot. (London),* 22, 513, 1958.
43. Chin, C., Haas, J. C., and Still, C. C., Growth and sugar uptake of excised root and callus of tomato, *Plant Sci. Lett.,* 21, 229, 1981.
44. Street, H. E., Growth in organized and unorganized systems — knowledge gained by culture of organs and tissue explants, in *Plant Physiology,* Vol. 5B, Steward, F. C., Ed., Academic Press, New York, 1969, 3.

Part II
Carbon Nutrition

Chapter 4

EFFICIENCY AND CAPACITY OF THE PHOTOSYNTHETIC LIGHT REACTIONS

Barbara A. Zilinskas

TABLE OF CONTENTS

I. INTRODUCTION

Photosynthesis is the process by which certain organisms, i.e., plants, algae, and selected bacteria, harvest solar energy and transduce it into the chemical energy of carbohydrate. The process involves transfer of electrons from a donor, i.e., H_2O in the case of higher plants and algae, to an acceptor, ultimately CO_2, through a series of charge transfer reactions. This review will concern selected aspects of the first of the steps leading to carbohydrate storage: the absorption of light by the photosynthetic pigments, transfer of this excitation energy to the photochemical reaction centers where primary energy conversion occurs, and transfer of electrons through a number of carriers resulting in production of ATP and NADPH. Emphasis will be placed on the potential for improvement of crop yield through change in the efficiency and/or capacity of the light-dependent photosynthetic reactions, based on examples of the adaptability and fine-tuned regulation of the photosynthetic light reactions. Discussion will be limited to higher plants, although the potential economic value of aquaculture should not be underestimated. For in-depth treatises in the photosynthetic light reactions, the reader is advised to consult several recent, excellent monographs.[1-3]

II. PHOTOSYNTHETIC MEMBRANE STRUCTURE AND FUNCTION

A. Pigments

All plants and algae contain chlorophyll *a* (chl *a*), a tetrapyrrole coordinated by magnesium in a ring structure. This chromophore is noncovalently bound to protein in the form of several pigment-protein complexes. This association with protein, in addition to solvent effects in its membrane environment, and its aggregation state, results in a variety of spectral forms of chlorophyll which in effect allows for more effective utilization of the solar spectrum.[4] Two specialized forms of chl *a*, called P_{680} and P_{700} for the wavelength maximum of the light-induced absorbance decrease upon photooxidation, are the reaction centers of photosynthesis where the actual photochemistry takes place. Other chl *a*-protein molecules comprise light-harvesting complexes which cooperate in the transfer of excitation energy to the reaction centers (see below). Each reaction center appears to be associated with 40 to 60 chl *a* accessory pigments in a core complex.[5-7] In addition to chl *a*, higher plants and green algae contain chlorophyll *b* (chl *b*) in the form of a chl *a/b* pigment protein complex, called light-harvesting complex (LHC), which denotes its sole function as an antenna unable to do photochemistry. It is the main light collector of mature plants.[7-8] Carotenoids are also light harvesters, albeit rather ineffective, with transfer to chl *a* of higher plants only 40% efficient;[9] they more importantly serve to protect chl *a* from irreversible photooxidation.[10]

Although the green algae closely resemble the higher plants in pigment composition, the other algal groups are widely divergent with regard to the light-harvesting apparatus. The common denominator is P_{680} and P_{700} reaction centers and a central core of chl *a* light-harvesting molecules. Additional light-harvesting capacity is met by a variety of pigment-proteins, including the water soluble phycobiliproteins of the blue-green and red algae, the peridinin chl *a* pigment-proteins of dinoflagellates, and the membrane-embedded fucoxanthin-chl a/c_2 proteins of the chromophyta, all of which efficiently transfer energy to the reaction centers.[11-13] In most cases, these pigment-proteins absorb light not effectively absorbed by chl *a* to enable these organisms to live in light-limiting environments where green algae or higher plants would not survive.

B. The Photosynthetic Unit

As described above, the photosynthetic pigments are of two major types, the specialized chl *a* reaction centers and the light-harvesting molecules. This presupposes the concept of the photosynthetic unit, a cooperation of pigments in collection of light energy and funneling of this energy to the reaction centers, where the actual photochemistry occurs. Emerson and Arnold[14] first showed that 2400 chl molecules are involved in the evolution of one O_2 molecule. As we now know that there is a transfer of four electrons twice in the photosynthetic light reactions, this suggests that each reaction center has a collection of 300 light-harvesting molecules associated with it. This number varies somewhat from species to species, within a species by environmental conditions, but basically remains between 200 and 400 light-harvesting pigments per reaction center.[15] The size of this antenna may be one of the limitations in photosynthetic efficiency, which shall be discussed in detail below. Also, recent evidence suggests that the original concept of the photosynthetic unit is too simplistic and rigid; these data and their implications follow below.

The two reaction centers serve as traps for two light reactions occurring in series.[16-17] Excitation energy is transferred via the light-harvesting pigments to the reaction centers where charge separation occurs; the reaction center is photooxidized and the primary acceptor reduced. The P_{680} reaction center serves as the trap for photosystem II (PS II) and will become reduced by electrons donated ultimately from H_2O. P_{700}, the center of photosystem I (PS I), similarly becomes photooxidized and is reduced by electrons passed from P_{680} through an intersystem electron transport chain connecting PS II and PS I in series; as a result of vectorial electron flow across the thylakoid membrane, ATP is produced. Electrons leaving P_{700} reduce NADP (noncyclic electron flow) or are recycled back to P_{700} through the intersystem chain (cyclic electron flow) to generate "extra" ATP. NADPH and ATP are then used to generate carbohydrate from CO_2.

C. Chloroplast Molecular and Structural Organization

Chloroplast thylakoids have five supramolecular complexes which participate in NADPH and ATP production. These are (1) the PS II complex containing P_{680} and its 40- to 60-antenna chl *a*, the immediate donors and acceptors of PS II, and the O_2-evolving machinery, (2) the cytochrome (cyt) b_6-f/Rieske Fe-S complex, comprising the bulk of the intersystem chain, (3) the PS I complex containing P_{700} and its 40- to 60-antenna chl *a*, subunit III and Fe-S centers, (4) the light-harvesting complex (LHC), which contains all the chl *b* and up to one half of the chl *a* of the chloroplasts, and (5) the ATP synthetase, which is made up of the intrinsic component CF_0 and the extrinsic CF_1, both of which are required to produce ATP at the expense of the proton gradient generated as a result of vectorial electron flow.[18,19] The structural identities of these integral supramolecular complexes are yet under active investigation, but tentative structure/function assignments have been made such that each complex corresponds to a distinct particle observed on freeze fracture of thylakoid membranes.[8,20-24]

All chloroplasts which contain the chl a/b (1:1) LHC contain appressed thylakoids organized into stacks or grana, as well as unappressed (unstacked) thylakoids; only the unappressed thylakoids and the grana end membranes and margins have contact with the stroma, and thus the unappressed thylakoids are often referred to as stroma thylakoids. The inner surfaces of the grana and stroma thylakoids enclose a continuous space between the two. Stroma and grana lamellae can be separated by detergent or mechanical shearing followed by fractionation by density gradient centrifugation[25] or polyethylene glycol-dextran phase partitioning.[26]

Analysis of these membranes, by freeze-fracture electron microscopy, lithium dodecyl sulfate polyacrylamide gel electrophoresis (for chl protein composition), and by measurements of the partial reactions of electron transport, provided a startling find. Nearly all the PS I and the CF_1-CF_0 supramolecular complexes are localized in stroma-exposed thylakoids, while all the PS II supramolecular complex is confined to the grana.[8,19,25-28] The localization of the cyt b/f complex is not clear; some reports suggest that the complex is equally distributed between appressed and unappressed membranes,[29-31] while other arguments support the exclusive localization of the complex in stroma-exposed thylakoids.[32-35]

The major portion of the chl a/b protein is found structurally associated with the PS II complex in the grana, and has been called LHC II. It is now clear that LHC II is responsible for grana stacking.[8] Kyle and co-workers[36] have provided evidence for a second pool of LHC II which is not tightly associated with the PS II complex but rather moves freely laterally in the membrane between grana and stroma lamellae and may thus participate in control of excitation energy distribution (see later). In addition, another light-harvesting antenna, LHC I, has been proposed which contains both chl a and b (4:1) associated with four polypeptides and the PS I core complex in the stroma-exposed thylakoids.[21,23,37] The proposed structural organization of the thylakoid membrane is schematically shown in Figure 1.

The unanticipated spatial separation of PS II and PS I centers raises some serious questions about the simplicity of the Hill and Bendall or "Z" scheme. Are there single electron transport chains connecting two reaction centers and their associated light-harvesting pigment-proteins? What is now meant by a photosynthetic unit, and how could there be a structural equivalent of this cooperation of light gatherers? How can light energy absorbed by the major light-harvesting unit, localized in the grana, be distributed equally to both photosystems? In short, with this structural arrangement, how can photosynthesis be as efficient as it is? Or, could it be possible that this spatial organization might be partly responsible for photosynthetic efficiency?

Since the PS II, PS I, and cyt b_6/f supramolecular complexes are structurally separated, there must be shuttles of electrons from one complex to the next. Plastoquinone, the lipid soluble electron transport component, transports electrons between the PS II and cytochrome complexes; this diffusion-limited lateral transport may account for the rate limitation of photosynthetic electron flow.[18] Plastoquinone exists as a pool of 10 molecules per P_{680} reaction center and is thought to link several reaction centers.[38-40] The second mobile electron transport carrier, plastocyanin, would, in turn, transfer electrons from the cyt b_6/f complex to PS I. Unlike plastoquinone, this carrier is water soluble and is therefore proposed to move along the inner thylakoid surface or through the locular space to P_{700}, localized in stroma-exposed membranes.[18,19,41,42] The last mobile carrier, ferredoxin, links PS I with the cyt b_6/f complex (in cyclic electron flow) or ferredoxin-NADP-oxidoreductase (in noncyclic flow to NADP).[18,19]

The implication following from such an organization of thylakoid electron transport components is that there is no fixed stoichiometry between the three supramolecular complexes participating in electron flow. The very existence of pools of linking electron transport intermediates, i.e., plastoquinone and plastocyanin, suggests that the integral membrane protein complexes may not occur in stoichiometric ratios. Again, these findings challenge the very meaning of the photosynthetic unit, if, in fact, there is no such entity as a single electron transport chain.[18,19] Melis and co-workers[43-47] reported stoichiometries of PS I to PS II reaction centers significantly departing from unity. While these findings have been confirmed in other laboratories,[48,49] a recent report has presented conflicting results.[50] In the work of Melis and co-workers,[43-47] PS I reaction center concentration was measured directly as the light-induced absorbance change at 700 nm, while the PS II reaction center concentration was approximated from the light-induced absorbance change at 325 nm, due to the reduction of the PS II primary acceptor Q. In addition to showing the localization of PS II and PS I reaction centers in grana and stroma lamellae, respectively, they showed that the ratio of 3.3 of PS II/PS I centers in developing pea changed to 0.77 in mature pea chloroplasts.[44] Other varia-

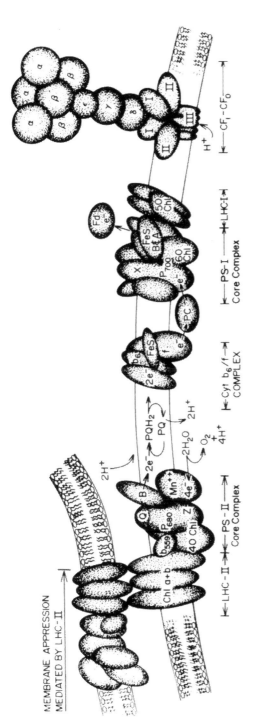

FIGURE 1. A model of the structural units of function which catalyze the light-harvesting electron transport, and energy coupling reactions of photosynthesis. PS II consists of a light harvesting pigment-protein complex (LHC II), a core complex containing the P_{680} reaction center, and a manganese-containing water splitting complex. LHC II mediates membrane appression and regulates excitation energy distribution. Electrons are transferred from water to plastoquinone (PQ); a cyt b_6-f Rieske Fe-S complex oxidizes PQH_2 and donates electrons to plastocyanin (PC). PS I, comprised of a light-harvesting pigment-protein complex (LHC I) and a core complex, oxidizes plastocyanin (PC) and transfers electrons to $NADP^+$. The ATPase, utilizing a proton gradient to drive ATP synthesis, consists of a hydrophobic CF_0 and an extrinsic CF_1. Model adapted from Kaplan and Arntzen.[8]

tions in electron transport carriers and reaction center stoichiometries appear to be more often the rule than the exception and will be used below to illustrate flexibility in photosynthetic membrane organization.

D. Regulation of Excitation Energy Distribution

This structural relationship of pigment systems and two light reactions also raises the question of how light energy is distributed equally to PS II and PS I to allow photosynthesis to proceed efficiently. As elegantly pointed out by Emerson and Lewis,[51] if light energy is funneled preferentially to one photosystem (as can be done by illumination with 680 nm light, absorbed almost entirely by PS I with the predominance of its long wavelength chl *a* spectral forms feeding P_{700}), photosynthesis does not proceed for lack of electrons to rereduce P_{700}; this is the classical "red drop" effect. A mechanism must exist to distribute excitation energy equally to PS I and PS II to maintain high quantum efficiencies.

Due to their different environments, PS II and PS I chlorophylls have different absorption properties. The chl *a* of PS I is enriched in long wavelength-absorbing forms (680 to 710 nm), while the shorter wavelength-absorbing forms (660 to 680 nm) are found in the PS II complex.[6,21] These chls comprise approximately 40 to 50% of the total chl of mature plants and the remainder is made up by the chl a/b LHC, which absorbs maximally between 650 and 680 nm.[52] Action spectra show that the LHC II primarily excites PS II,[53] and structural data indicate that this complex is localized in the grana.[8,20,24,54] Seventy percent of the light absorbed by the chloroplast is absorbed in the grana, while the PS I-rich stroma lamellae receive only about 30% of the initial quanta trapped by the chloroplast. As efficiency of noncyclic electron flow requires that both reaction centers turn over at the same rate, excitation energy must somehow be evenly distributed to the two reaction centers. If PS I were to receive less excitation energy, it would operate more slowly than PS II and the overall process would become limited. A mechanism must exist which allows equal distribution of excitation energy such that both photosystems operate in synchrony and efficiently. For some time, it has been known that the pigment bed undergoes adaptive changes which correct over-excitation of one pigment system or the other.[55,56] Only recently have biochemical mechanisms been proposed and adequately tested to explain how these adaptive changes are brought about. Several detailed reviews[57-59] have recently appeared which address this subject, but a brief summary will be presented here.

Bennett[60-62] showed that a membrane-bound phosphatase and kinase could catalyze the reversible phosphorylation of threonine residues on part of the LHC II exposed to the stroma. The phosphorylation depended on the activation of the kinase which in turn occurred only when the plastoquinone pool was reduced; dephosphorylation by the phosphatase resulted when the kinase was inactivated following reoxidation of the plastoquinone pool.[63,64] The significance of the reversible phosphorylation rests in its resulting in alterations in excitation energy distribution. Phosphorylated LHC II transfers excitation energy preferentially to PS I; dephosphorylation results in redirection of energy flow to PS II.[65] As phosphorylation/dephosphorylation reflects the redox state of the plastoquinone pool, this is a logical mechanism to ensure that neither photosystem operates faster than the other, an effective check-and-balance system.

Very recently, it has been shown that there is a mobile pool of LHC II which, when phosphorylated, moves from the grana to the stroma lamellae, thereby allowing directional flow of excitation energy to PS I. Dephosphorylation of LHC II, following reoxidation of the plastoquinone pool, results in return of this pool of mobile LHC I to the grana lamellae, once again directing energy transfer from LHC II to the PS II centers.[36]

An explanation for the separation of PS I and PS II into stroma and grana lamellae, respectively, has been recently offered by Staehelin and Arntzen.[59] There must be a mechanism to prevent all the excitation energy harvested by any of the antenna systems from being all drained into PS I, since this pigment-protein complex has the longest wavelength absorption and is more efficient than PS II in trapping.[66] To prevent this excessive funneling of excitation energy to PS I, the two photosystems have been spatially segregated.[46] Moreover, to correct for imbalances of energy distribution to each photosystem, controls via a mobile light-harvesting unit exist.[36] A more elegant system would be difficult to imagine.

III. LIMITATIONS OF THE LIGHT REACTIONS

A. Efficiency of Light Utilization

Photosynthesis is the most efficient known system for the conversion of solar energy into chemical energy; yet, needless to say, there is always room for improvement. One of the most controversial debates in science concerned the maximum quantum yield in photosynthesis. Using manometry to measure O_2 and CO_2 exchange of algae, Warburg and co-workers insisted that the minimum quantum requirement (1/O) was 4 hν/O_2, which implies that for every four photons absorbed, four electrons are transferred and a molecule of O_2 evolved. With 700 nm light, the quantum efficiency of energy conversion would be ∼70%, which approaches the thermodynamic limit of 75%. However, Emerson and co-workers (and many others that followed) showed that the minimum quantum requirement was not 4 but 8 to 10 hν/O_2 (see Kok[67] for a critical discussion of the quantum yield debate), reducing the quantum efficiency to a more realistic value of 30 to 35%. The establishment of a minimum quantum requirement ≥8 provided the first suggestion of two photoreactions which cooperate in series in the evolution of O_2 from H_2O.

Of course, efficiency of conversion of sunlight into carbohydrate in the plant under field conditions never approaches 30%. Bolton[68] and Bolton and Hall[69] described efficiency for photosynthesis and reasons for loss of energy in solar conversion, which are summarized as follows. First, quantum efficiency is that fraction of absorbed quanta which is utilized. Forty-seven percent of sunlight energy falling on a leaf is lost because the radiation lies outside the photosynthetically active region (360 to 720 nm). Also, within the region of photosynthetically active radiation, generally, another 30% of this incident energy is lost due to incomplete absorption by chloroplast pigments or by absorption by nonphotosynthetic components in the leaf. Of the energy then actually absorbed by the chloroplast pigments, 24% of this is, in turn, wasted by internal conversion or degradation of the excess energy of the absorbed photons to the excitation energy at λ max, 700 nm. Of the 28% of incident solar energy remaining, 68% of this is lost during the conversion of the excitation energy at 700 nm to the chemical energy stored in glucose, which reduces the total energy remaining to 9%. Dark respiration and photorespiration account for further reduction in the efficiency of energy conserved from incident sunlight, yielding a final net efficiency of 5 to 6%. Under field conditions, further losses might be expected even in the most intensive agricultural systems (supplied with adequate H_2O, nitrogen, etc.). Maximum field efficiencies for C_3 and C_4 plants under optimum conditions are now reaching levels of 2 and 4% efficiency, respectively.[70] If one measures efficiency of crop photosynthesis over a yearly basis, then most crop plants in most geographical areas would be far less efficient than the numbers reported above. Table 1 illustrates the maximum photosynthetic efficiency of a maize crop, as described earlier by Ku and Edwards,[71] based on data of Loomis and Williams[72] and Zelitch.[73]

B. Efficiency and Capacity

The light saturation curve of photosynthesis provides two critical pieces of infor-

Table 1

AN EXAMPLE OF MAXIMUM PHOTOSYNTHETIC EFFICIENCY OF A MAIZE CROP[a]

Average daily total solar radiation	500 cal/cm²·day
PAR (400—700 nm)	222 cal/cm²·day
Total quanta available, 400—700 nm (using value of 8.64 μmol of quanta/calorie of total solar radiation	4320 μmol quanta/cm²·day
Loss by reflection (8%)	−360 μmol quanta/cm²·day
Total quanta absorbed	3960 μmol quanta/cm²·day
Assume quantum efficiency = 1 CO_2 fixed/17 quanta absorbed = 1 μmol CO_2 fixed/17 μmol quanta. Then μmol CO_2 fixed/3960 μmol quanta =	233 μmol CO_2/cm²·day
Respiratory loss (estimated at about one third of photosynthesis)	278 μmol CO_2/cm²· day
Net production of carbohydrate (CO_2 fixed)	155 μmol CO_2/cm²·day
1 mol CO_2 fixed to level of sucrose has 120 kcal energy, or 1 μmol CO_2 converted to sucrose = 0.12 cal. Chemical energy stored/day = 0.12 (155) =	18.6 cal/cm²·day
Efficiency of use of total incident radiation =	$\dfrac{18.6 \text{ cal/cm}^2 \cdot \text{day } (100)}{500 \text{ cal/cm}^2 \cdot \text{day}} = 3.7\%$

[a] Table taken from Ku and Edwards,[71] based on data of Loomis and Williams[72] and Zelitch.[73]

mation. At very low light intensities, the linear part of the curve indicates the efficiency of photosynthesis; the quantum yield of the reaction is defined by the slope of the curve, and the quantum requirement is the reciprocal of that value. In isolated thylakoids, the light-independent region at saturation reflects the capacity or number of active electron transport chains. This is so because at light saturation the rate is limited by the dark reactions, which, in the case of the Hill reaction in isolated thylakoids, is the turnover time of the electron transport chain, limited by the slowest step in electron transport. In leaves of C_3 plants in air, light saturation most likely occurs because of CO_2 availability[74] and the rate limitations in CO_2 fixation by ribulose bisphosphate (RuBP) carboxylase[75] (however, see Section IV). In contrast to C_3 plants, C_4 plants normally do not show light saturation, even at full sunlight.[76] However, the quantum efficiency of C_3 photosynthesis at lower light intensities and reduced O_2 tension is considerably greater than that of C_4 plants because of the extra expense of $2ATP/CO_2$ fixed in the C_4 pathway[71] (see Chapter 6, Volume I). Nonetheless, in C_3 plants, as leaves in the upper crop canopy become light saturated, absorbed energy is dissipated as heat, and the quantum efficiency decreases. This loss results either when CO_2 is limiting or when the level of absorbed energy exceeds the photochemical capacity of the electron transport chain.

As a consequence of the photosynthetic unit, light of widely different wavelengths and low intensity can drive photosynthesis. At one tenth full sunlight, a cloudy day, each reaction center receives about 300 excitons per second from its associated 300 light-harvesting molecules. To maintain a high quantum efficiency even at this reduced light intensity, the electron transport chain must turn over at least 300 times per second. At full sunlight, the turnover rate must be ten times that to keep pace with the excitation trapping rate. In actuality, the half-time for the oxidation of plastoquinone is 20 msec, which then becomes the rate limiting step in the light reactions.[38,77,78] Consequently, the maximum rates of the light reactions at saturating light intensities are usually less than 5% of those theoretically possible. There are two obvious solutions to the dilemma: either the turnover time at the rate limiting electron transport step must increase or the size of the photosynthetic unit (i.e., number of antenna per reaction center) should decrease. Moreover, this mismatch of light-harvesting capability and electron transport may only be the case for leaves near the top of the canopy receiving an excess of incident quanta, but not necessarily for shaded leaves toward the bottom of the canopy. Can one hope to improve capacity and/or efficiency of photosynthesis by genetic modification of the light-dependent reactions?

C. Sun-Shade Adaptation

Several investigators have examined the flexibility of the photosynthetic unit size, utilizing two basic approaches: environmental adaptation to light intensity and the use of chlorophyll-deficient mutants. It is well established that "shade" plants are able to efficiently use low intensity light for photosynthesis, but photosynthesize poorly, if at all, at high light intensities; in contrast, "sun" plants have a high capacity for photosynthesis at high light intensities but have low rates of net photosynthesis at low intensities.[79-85] Initial slopes of the light response curve, indicative of the quantum yield for electron transport, remain constant among both sun and shade plants.[79,83] The basic adaptability or potential for such environmentally induced change is determined by the genotype, with genotypically determined adaptability existing both between different species as well as between different ecotypes within a species growing in shaded or sunny habitats. Several physiological parameters affecting photosynthetic efficiency and capacity appear to be altered by light intensity adaptation. These include: leaf anatomy, stomatal conductance, chloroplast number and structure, soluble protein content — particularly RuBP carboxylase amounts, and pigment and electron transport carrier content.[78-87] Therefore, it is difficult to dissect out contributions of individual alterations to change in total net photosynthesis. Light adaptation is probably not achieved by changing a single rate limitation, but by balanced changes in the capacities of potentially several rate limiting steps. Two points seem clear, however; namely, that increased capacity for whole leaf photosynthesis in sun plants is correlated with an increased amount and activity of RuBP carboxylase[83,89] as well as increased light-saturated Hill reaction rates in thylakoids obtained from sun leaves.[79,80,89] The following discussion will focus on the reasons why the light-saturated rates of photosynthetic partial reactions (based either on chl or leaf area) are several fold greater in sun than in shade plants.

Shade plants, such as those obtained from rainforest floor habitats, have low protein/chl ratios (most importantly due to decreased level of RuBP carboxylase), decreased chl a/b ratios, well developed grana (due to an increase in LHC II), and an increase in total chl/leaf area.[19,83] These observations suggest that shade plants have a larger photosynthetic unit size than sun plants (that is, more antenna/reaction center), allowing them to better utilize the limited light available in the densely shaded habitat. While rates of both PS I and PS II partial reactions at rate-limiting light intensities (quantum yields) are similar in sun and shade species,[79,83] there are remarkable differences (10- to 15-fold higher PS II activities and 6-fold higher PS I activities) in light-saturated rates.[79,80,89] As rates are reported on a chl basis, the low rates of shade plant chloroplast reactions are due to either an increased photosynthetic unit size or a decreased maximum rate of electron transport.

Some controversy exists in the literature concerning the reason for the reduction in photosynthesis rates in shade plants. One argument favors increased photosynthetic unit size, with larger but fewer photosynthetic units per leaf area.[82,90-94] A second explanation is that there is little change in antenna size/reaction center, but there are sizeable differences in capacity resulting from decreased levels of electron transport components in shade plants.[45,47,79,80,83,89,95] The consequence of either is a reduction in ATP and NADPH production and a subsequent lowering of CO_2 fixation rates.

Most of the seeming contradiction can be resolved in the light of recent redefinitions of the "photosynthetic unit" (see Section II.C). Until recently, the photosynthetic unit was imagined as an electron transport chain with 1 P_{700}:1 P_{680}:1 cytochrome f and antennae pigments supplying each reaction center. Therefore, when values of total chl to P_{700} were reported, it was assumed that this ratio also reflected that of total chl/P_{680}

and total chl/cyt f. Therefore, increased total chl/P_{700} ratios were interpreted as the result of larger photosynthetic unit sizes.

There is, however, growing evidence for unequal stoichiometries of both reaction centers[44,49] and electron transport carriers.[19,79,80,83] While it has been known for some time that there are large changes in LHC II content in thylakoids of shade plants, unexpectedly there is a parallel increase in PS II reaction centers.[45,47] This, in effect, results in a constant functional antenna size for PS II. Likewise, there is no change in the functional antenna size for PS I.[45,47,95] Reporting total chl/P_{700} changes as photosynthetic unit changes would be misleading, as, in fact, P_{680}/P_{700} stoichiometries are changing. Plants taken from their natural shade habitats have P_{680}/P_{700} ratios as high as 3.9.[45] An increase in the PS II/PS I reaction center ratio (with an unaltered antenna size for each) would increase the total chl/P_{700} ratio. In shade plants then, it appears that proportionately more chl is associated with PS II, both in numbers of PS II reaction centers and the PS II associated antenna, LHC II. This adjustment is in compensation for the far-red enriched light conditions usually experienced in nature by shade plants, offsetting the imbalance created by far-red illumination which primarily excites PS I.[45] While PS II centers increase in number, the number of intersystem electron transport carriers remains small in shade plants, and, as a consequence, the total capacity of photosynthetic electron flow is limited. Adaptation to sun conditions results in a readjustment of stoichiometry of reaction centers and associated antenna (antenna size/reaction center remaining constant) and addition of "extra" intersystem electron transport carriers and ATP synthetase.[97]

These genotypic adaptations enable plants to photosynthesize efficiently under the extreme environmental conditions they experience, but it precludes high efficiency at the other environmental extreme.[83,84] As Björkman[98] has pointed out, it is questionable whether efficiency can be improved over that which has evolved naturally to environmental stress. Nonetheless, examination of the underlying mechanisms and limits of adaptation may point to potential avenues for deliberate change; for example, based on sun-shade adaptation, one might conclude that it may be easier to alter relative stoichiometries of reaction center, cyt b_6-f, or CF_1-CF_0 complexes to effect increased photosynthetic capacity than to adjust antenna size/reaction center.

D. Chlorophyll-Deficient Mutants

Many types of mutations at the thylakoid level have been reported, but all of those mutations affecting electron transport components have been lethal. Levine[99] and co-workers effectively utilized these mutations occurring in *Chlamydomonas reinhardtii* to establish the ordering of electron carriers in the electron transport chain.

Although most chl-deficient mutants are lethal or grow very slowly,[100] several have been described which are as photosynthetically capable as the wild type and in some cases have electron transport rates (on a chl basis) that are several fold higher than those of the wild type.[101-110] Chlorophyll-deficient mutants have been used to examine possible correlations between chloroplast composition and structure and photosynthetic units.

Several possible changes can occur that result in an overall chl deficiency. These include: a reduced number of photosynthetic units without alteration in unit size, a reduced photosynthetic unit size, an increased unit size with a concomitant decrease in the number of units per leaf area, and a decreased unit size and number. Examples of each of these have been provided in the last 15 years, with the most commonly described mutant type being that which has smaller photosynthetic units.[15,103,111,112] These mutants typically have a decrease in the relative amount of LHC II, measured as increased chl a/b ratios and decreased amounts of LHC II on mildly dissociating SDS-PAGE gels, and reduced grana formation. In addition, they have decreased chl/P_{700} ratios, require higher light intensities for saturation, and demonstrate considerably higher rates of both PS I and PS II reactions on a chl basis.[80,113,114] These characteristics are reminiscent of the features of sun-adapted plants, and consequently, there is some skepticism on the interpretation of the data as simple modification of photosynthetic unit size. However, in the case of these chl-deficient mutants unlike the situation in sun-shade adaptation, measurements of the effective absorption cross-sections of each reaction center indicate that, at least in the tobacco mutants studied, the antenna sizes for P_{700} and P_{680} were decreased;[43] in addition, the mutants showed an increase in the stoichiometric ratio of PS II/PS I reaction centers.[43,44] This result has recently been supported by additional work in two other laboratories.[48,115] In cases of extreme deficiency such as seen in the tobacco mutants, the size of the antenna associated with PS II and PS I seems to be essentially that of the PS II and PS I core complex themselves,[43] suggesting the loss of LHC II and LHC I (both of these complexes containing chl *b* as components). But all ranges of chl deficiency have been seen in the mutants examined. Mutant barley retains 70% of its chl,[101] pea has 50%,[103] and the tobacco Su/su mutant has only 20 to 30% of the chl of the wild type.[102] CO_2 fixation rates expressed on a chl basis at light saturation showed an inverse relationship with leaf chl content; wild type barley, pea, and tobacco had CO_2 fixation rates 70%, 46%, and 24% of the rates of the respective mutants.[103,116,117] The higher intensities required to achieve saturation were explained by a smaller photosynthetic unit size, although chl distribution in each antenna system has only recently been investigated and only in the tobacco mutants.[43,48] Unlike the sun-shade plants, the quantum efficiencies of the mutants at low light intensities were lower than those of the wild type.[80,103,117] As the mutants, on a leaf area basis, had photosynthetic rates that were roughly equivalent to those of the wild type at saturation intensities while rates on a chl basis were considerably higher in the mutants, it appears that there is a decrease of unit size in these mutants with no alteration in the number of photosynthetic units per leaf area. To achieve higher leaf photosynthetic rates, the ideal situation would be met by a mutant with smaller photosynthetic units and concomitantly a larger number of units per leaf area. This modification would provide an advantage to plants occupying sunny habitats.

Several other chl deficient mutants have been described which depart from the usual characteristics and are worth mentioning as they illustrate the variety of possible mutant phenotypes and provide suggestions for future deliberate genetic alteration. One of these is a virescent mutant of cotton.[107,111,118] At the early stages of development, this mutant has three- to fivefold less chl than the wild type and a three- to fivefold higher photosynthetic rate on a chl basis, both at low and high intensities. Moreover, the CO_2 fixation rate on a leaf area basis (at low and high intensities) and chl a/b ratios are nearly the same in the mutant and wild type. Light saturation is reached at the same intensity in mutant and normal plants. With maturity, the yellow leaves green and are indistinguishable from the wild type.

Nielsen et al.[109] examined several barley mutants which had lowered total chl/leaf area and either higher chl a/b or lower chl a/b ratios than the wild type. Those with low chl a/b ratios had extensive grana formation and lower rates of photosynthesis on a chl basis than the wild type; the reverse was seen with the mutants having high chl a/b ratios. These results support the hypothesis of Anderson[18,19] that, coincident with larger amounts of exposed thylakoids (relative to thylakoids appressed in grana regions), there are a greater number of those membrane complexes proposed to be localized in exposed membranes (cyt b_6-f complex, plastocyanin, PS I complex, ferredoxin-NADP oxidoreductase, and ATPase); this allows for an increased production of NADPH and ATP under light-saturation conditions, afforded by the higher capacity of electron transport carriers and associated enzymes.

Alberti et al.[119] examined a virescent mutant of peanut and found that its lower rates of photosynthesis both on a leaf area and chl basis could be accounted for by a larger relative decrease in the PS I core complex (and presumably P_{700} reaction centers) than in the LHC II chl a/b protein. This study[119] and others,[95,112,115] where conclusions are based primarily on the analysis of the relative amount of chl associated with chl-protein complexes obtained by mildly dissociating SDS-PAGE, unfortunately do not provide actual measurements of P_{680} or P_{700} and, consequently, conclusions are provisional.

A soybean mutant has been described[104,105] which presents a potentially "useful" altered genotype. This mutant has a 2.5-fold decrease in total chl per leaf or gram fresh weight and a somewhat smaller decrease in total protein; there is a 2.5-fold increase in the chl a/b ratio accompanied by reduced grana formation. Chl/P_{700} ratios are identical in the soybean mutant and wild type, and concentrations of cytochromes (as well as plastoquinone) on a chl basis are increased in the mutant. Light-saturated rates of electron transport were increased by three- to fivefold in the mutant. However, unlike most chl-deficient mutants, light saturation occurred at similar intensities for both mutant and wild type soybeans. The quantum yields of photosynthesis extrapolated to zero intensity were the same in both genotypes, and at finite light intensity, the mutant actually showed an increased quantum efficiency. The rates and extent of plastoquinone reduction and oxidation were measured in both mutant and wild type. The mutant had a twofold greater apparent first order rate constant for plastoquinone oxidation and a three- to fivefold larger pool of readily reducible plastoquinone. This is an important finding; as the oxidation of plastoquinone is known to be the rate limiting step in electron transport,[38,77] increases in photosynthetic rates in the soybean mutant can probably be attributed to a faster turnover time of the electron transport chain. It is interesting to note that analysis of lipid composition of the soybean thylakoid membranes showed quantitative and qualitative changes; in the light of our knowledge that plastoquinone serves as a shuttle in the lipid bilayer of the thylakoid, one might speculate that alteration in the chloroplast lipids may affect the rate limiting step in photosynthetic electron transport.

E. Do the Light Reactions Limit Photosynthesis?

Until recently, limitations in photosynthesis were thought to reside primarily in CO_2 fixation or the dark reactions of photosynthesis.[120-123] These limitations included the low (300 ppm) atmospheric CO_2 concentration, diffusional barriers of CO_2 to the chloroplast,[74] and problems inherent to RuBP carboxylase-oxygenase.[76,124,125] Convincing experiments included CO_2 enrichment studies, where doubling atmospheric CO_2 concentrations resulted in a doubling of intensities at which light saturation was reached with a resultant twofold increase in leaf photosynthetic rates.[126] As a consequence, efforts to improve photosynthesis by breeding or nonconventional genetic means have largely been directed toward events dealing with CO_2 utilization.[123,127]

Recent findings, however, suggest that the light reactions may at least co-limit photosynthesis. Utilizing iron deficiency as a means to specifically and quantitatively decrease thylakoid content of the leaf, Terry and colleagues found that the rate of leaf photosynthesis was directly related to the amount of thylakoids and electron transport activity.[128,129]

The normal chl content of leaves is 400 to 600 mg chl m^{-2}, and these leaves absorb between 80 and 85% of sunlight in the wavelength range of 400 to 700 nm. Increases in chl content from 250 to 750 mg m^{-2} result in significant increases in absorption only in the green and far-red spectral regions where chl absorbs poorly (for example, at 550 nm from 60% to 82% absorption at the respective leaf chl contents above).[83] This increase would only be advantageous under very low light intensities where photosynthetic rates are light limited.

It is not surprising, then, that when chl content is reduced by 90% by severe iron depletion, absorption by the leaves is reduced by only 30%. Moreover, iron deficiency has no effect on the photochemical conversion of the absorbed light, if these quantum yields are properly measured at very low light intensities.[130] As with sun-shade adaptation, photosynthetic capacity rather than efficiency is adversely affected by iron stress. Thus, differences in photosynthetic rates exist between iron-sufficient and deficient plants at high light intensities and light saturation is reached at lower intensities in iron stressed plants, while there is no difference in electron transport rates or whole leaf photosynthesis as measured by gas exchange at low light intensities.[130] The advantage offered by the iron stress experimental system is that the only parameter altered is at the thylakoid level; no changes occur in Calvin cycle enzymes,[130,131] chloroplast number per cell or per leaf area, number of cells per unit leaf area, chloroplast replication, leaf thickness, leaf fresh weight per area,[130,132] stomatal conductance, CO_2 compensation point, or dark respiration.[128] Therefore, it provides an excellent opportunity to examine the effects and possible limitations of photochemical capacity on total leaf photosynthesis.

When rapidly growing plants are deprived of iron, the chloroplasts appear to be the only organelle affected.[133] The number of thylakoids per chloroplast is reduced, proportional to the degree of iron stress,[133] accompanied by a quantitative loss of all thylakoid components including chl *a* and *b*, carotenoids, electron transport components, membrane lipids and proteins, and lamellar iron and manganese.[129,134,135] Although some components are reduced in somewhat greater quantity than others, for the most part, iron stress results in a coordinate loss of total thylakoid constituents, while the enzymatic activity of CO_2 fixation is not altered. As leaf growth and chloroplast replication occur at the same rate in iron deficient and control plants, and as chl content per leaf does not change, it appears that thylakoids are conserved but diluted by leaf growth during deprivation; thylakoid content is restored to normal in four days upon iron resupply.[129,135]

Terry has compared gas exchange rates of leaves from iron sufficient and deficient plants at various CO_2 concentrations and light intensities and concluded that the light reactions co-limit total leaf photosynthesis at normal atmospheric CO_2 concentrations. At ambient CO_2 levels of 300 to 1000 ppm, photosynthesis/leaf area at light saturation increased almost linearly with an increase in thylakoid components and in vitro electron transport activity/leaf area. It is most interesting to note that at an ambient CO_2 concentration of 300 ppm (that occurring under field conditions) and at saturating light intensities, photosynthesis/leaf area increased by 36% with an increase of chl (a measure of total thylakoid components) from 400 to 650 mg m^{-2}, the range of chl concentration existing in iron sufficient control plants.[128,129] At saturating light, these increases must clearly reflect increases in photochemical conversion and not light absorption. However, even at nonsaturating light intensities, the increases in total photosynthesis/leaf area cannot be accounted for solely by increased absorption of light. In fact, even at the lowest irradiances, the major effect is on photochemical conversion.[128,129]

Several other previous studies have suggested that photosynthesis/leaf area is positively correlated with total chl content,[136-139] although others have refuted such claims either on a theoretical basis[140] or by observation of a lack of correlation between light absorption or chl content with photosynthesis/leaf area in a number of crop species.[121,141-143] This dispute, as well as support for the conclusions of Terry and co-workers,[128,135] might well be reconciled by realization that chl concentrations alone may not necessarily reflect reaction center and/or electron transport component concentrations, and one might expect these to be of paramount importance in rates of photosynthesis. Therefore, the suggestion offered by Terry[129] that breeding programs screen for increased chl leaf content as an index of increased photosynthetic capacity may not necessarily be an effective indicator of total thylakoid content in most crop species, as it appears to be in that crop studied by Terry, sugar beets.

A number of other studies have been done to assess the possibility that the light reactions limit photosynthesis. Lilley and Walker[144] showed that the activity of extracted RuBP carboxylase under optimum conditions with saturating RuBP was significantly higher than that seen in intact chloroplasts at nonlimiting CO_2 concentrations. Moreover, they showed that a fairly low concentration of CO_2 was needed for half-maximal photosynthetic rates, indicating saturation imposed by limitations in energy supplied by the light reactions; likewise, a departure from linearity of *in organello* photosynthesis with increasing CO_2 concentrations was interpreted as a limitation in ATP and NADPH generated in the light reactions. However, at CO_2 concentrations expected in nature, the capacity for CO_2 uptake with the intact chloroplast or with exogenous RuBP supplied to the isolated enzyme was approximately the same.[144] Collatz[145] reached similar conclusions by studying RuBP pool sizes in vivo. Von-Caemmerer and Farquhar[146] showed that CO_2 assimilation rates were limited by the RuBP-saturated rate of RuBP carboxylase-oxygenase at low intracellular CO_2 and by the rate of RuBP regeneration at higher CO_2 intracellular concentrations (limited by electron transport capacity). Under normal physiological conditions, the capacities of the two reactions (i.e., RuBP carboxylase activity and supply of reducing power) are approximately equivalent; the point of equivalence between light reaction and carbon metabolism capacity occurred at intracellular CO_2 concentrations of 250 μbar. Clearly then, if a concerted effort is being made to reduce some of the limitations in carbon metabolism, it appears that it would be shortsighted to not also consider possibilities for improving the efficiency and capacity of the photosynthetic light reactions.

IV. PHOTOSYNTHESIS AND CROP YIELD

Although one would intuitively assume that greater leaf photosynthesis rates would lead to higher crop yields, this is not necessarily the case, as reviewed recently by Gifford and colleagues[121,147] and Ellmore.[148] Gifford and Jenkins[121] have argued that crop physiology is a hierarchical process with limited likelihood of increasing crop yield by removing constraints at low levels of organization (e.g., thylakoid activity); rather, breeding efforts should be directed at high levels of organization, i.e., the crop canopy. To support their case, they cite numerous studies[149-158] where there is little positive correlation between yield and leaf photosynthesis rate.

The most frequently cited argument for the lack of correlation between light reaction capacity and crop yield is in studies of chl content.[140-143] It is clear, however, that leaf chl content alone will probably not be a good index of electron transport capacity which is most likely to provide the limitation to photosynthesis (see previous discussion). Attempts to correlate other characteristics of the photochemical apparatus with crop yield have been limited; assays are more cumbersome or require equipment. When electron transport or phosphorylation has been measured to compare different varieties or genotypes, most often rates are reported on a per chlorophyll basis without knowledge of chl per unit leaf area; these data cannot be usefully extrapolated to predictions of crop photosynthetic rates, much less crop productivity. Most breeding efforts have thus been made to improve harvest index, and consequently carbon partitioning is of critical importance.

Yet, studies of light enrichment show that crop yield is photosynthetically limited. Maize crops provided with a "light rich" environment showed significantly higher grain yields (as high as 80% increases when light-supplemented border rows were compared with densely populated rows).[159] Similarly, seasonal variation in solar radiation correlated with crop yields.[160] These results might be expected for maize, a C_4 plant, which does not reach light saturation even at full sunlight.[161] However, similar increases in soybean yield have been reported with supplemental light, with the largest increase (30%) found by supplemental illumination of leaves at the bottom of the canopy.[162] Similarly, selective defoliation[163] and shading[122,164] studies have resulted in the same conclusion that light can be a limitation to crop productivity.

A recent study provides an example for one possible explanation for the apparent paradox of limited correlation between leaf photosynthesis rates and crop yield. Wells et al.[165] noted that differences in leaf photosynthesis among soybean genotypes which have not correlated with crop yield probably resulted because the individual leaves measured were not representative of the soybean canopy. When they measured canopy-apparent photosynthesis, positive correlations with seed yield were seen,[165] in contrast to negative correlations with individual leaf photosynthesis of the same genotypes.[155] As it is clear that total assimilatory capacity of the soybean crop is better represented by canopy-apparent photosynthesis, this may be a good index for selection in breeding programs. Difficulties with individual leaf measurements may have arisen because photosynthesis varies with leaf age and position on the plant as well as environmental effects, and measurements of the entire canopy over an extended time period eliminate some of these problems.

Christy and Porter[164] also studied the relationship between canopy photosynthesis and yield in field-grown soybeans and found that grain yield strongly depended on seasonal photosynthesis. They cited several reasons for the reported lack of correlation between leaf photosynthesis rates and crop yield; first, the obvious is that photosynthesis rates are generally reported for individual leaves, while yield measurements are made on the entire canopy. Similarly, photosynthesis measurements are point-in-time determinations, made at different developmental stages of the plant, consequently not taking the entire growing season into account, as is the case with crop yield measurements. Also, the relationship between photosynthesis and yield is necessarily complex, differing from crop to crop and from cultivar to cultivar. Thus, there are a number of reasons to explain the seeming lack of correlation between individual leaf photosynthesis rates and crop yield. Christy and Porter[164] have introduced the use of "photosynthetic conversion efficiency" (PCE) which is the efficiency with which crops convert photosynthesis into grain yield (i.e., seasonal photosynthesis x photosynthetic conversion = crop yield efficiency). As Christy and Porter indicate, the term PCE differs from harvest index, as PCE is based on seasonal photosynthesis, while harvest index is a point-in-time measurement. Use of PCE will likely resolve the paradoxical problems relating photosynthesis and crop yields.

Ogren[127] has provided an additional explanation for the negative correlation sometimes reported between leaf photosynthesis and crop productivity. In soybean, cultivars showing high photosynthesis rates per leaf area generally have thicker leaves (with more photosynthetic machinery), while on a per tissue weight basis, rates of photosynthesis are nearly identical in all cultivars. In a crop situation, it is critical that there be early canopy closure to maximize seasonal photosynthesis and yield; thus, growth rate, especially the initiation and expansion of leaves, must be fast. This may necessitate the production of thinner and a larger number of leaves (with lower photosynthesis rates per leaf area), which would fill the canopy faster to maximize light interception.

Discussions to this point on light utilization have been largely limited to the level of chloroplast. Obviously, light interception depends on a number of other complex factors. Maximization of light interception by the crop can be met by (1) maintaining crops over land surfaces for as extended a time period as possible (leaf canopy persistence); (2) optimizing leaf surface area per unit ground during the crop growth cycle (leaf area index); and (3) providing an effective leaf array (orientation and angle) in the canopy (canopy architecture).[121] Although each of these parameters affects light interception and net canopy photosynthesis, they fall outside of the scope of this book, and the reader is thus referred elsewhere.[121,147] Nonetheless, in summary, it is clear that crop productivity is limited by many factors, including the interception and utilization of solar radiation.

V. SUMMARY

Although a thorough understanding of the mechanisms of the photosynthetic light reactions is yet to be achieved, discussion and experimentation are well underway to see how photosynthesis might be modified to improve its efficiency and capacity. Although there has been some debate on the validity of assuming a direct relationship between photosynthesis rates and crop yields, the rationale behind these efforts is that improvements in photosynthesis will lead to increased food production. This chapter has dealt with the prospects of increasing net photosynthesis by increasing the efficiency and/or capacity of the light reactions. Evidence has been presented which indicates that, contrary to popular belief, the availability of ATP and NADPH, produced by the light reactions, may co-limit photosynthesis. The photosynthetic membrane is adaptable, flexible, and influenced by the environment. To deliberately introduce alterations, such as increased electron transport capacity, beyond that which is possible in nature, may be difficult until more is known about the regulatory aspects of gene expression. Although many of the genes coding for chloroplast thylakoid proteins have recently been identified, significant barriers to genetically engineering a plant with improved photosynthesis must be overcome. Challenges include transformation of the chloroplast, control of expression of introduced genes, and integration of introduced genes into the already well-coordinated effort of nuclear and chloroplast genomes. As high as these hurdles may seem, examples of promise exist. McIntosh[166] and co-workers have recently shown that a single nucleotide substitution in the chloroplast gene coding for one of the intersystem electron transport components results in resistance to atrazine in *Amaranthus;* the next obvious step is to determine whether atrazine resistance can be transferred to a susceptible plant by gene transfer. With combined efforts of traditional plant breeding and nonconventional genetic manipulation, we shall soon see whether photosynthesis can be improved, hopefully resulting in increases in crop yields.

ACKNOWLEDGMENTS

New Jersey Agricultural Experiment Station, Publication No. F-01104-1-83, was supported in part by State funds and by the United States Hatch Act. This work was also supported in part by the Science and Education Administration of the U.S. Department of Agriculture under Grant 5901-0410-8-0185-0 from the Competitive Research Grants Office.

I thank Dr. C. J. Arntzen and Academic Press for providing Figure 1, and Drs. S. B. Ku, G. E. Edwards, and CRC Press for Table 1.

REFERENCES

1. Govindjee, *Photosynthesis, Vol. 1, Energy Conversion by Plants and Bacteria,* Academic Press, New York, 1982.
2. Clayton, R. K., *Photosynthesis: Physical Mechanisms and Chemical Patterns,* Cambridge University Press, Cambridge, England, 1980.
3. Barber, J., *Electron Transport and Photophosphorylation,* Elsevier, Amsterdam, 1982.
4. Sauer, K., Primary events and the trapping of energy, in *Bioenergetics of Photosynthesis,* Govindjee, Ed., Academic Press, New York, 1975, chap. 3.
5. Satoh, K., Polypeptide composition of the purified photosystem II pigment-protein complex from spinach, *Biochim. Biophys. Acta,* 546, 84, 1979.
6. Diner, B. A. and Wollman, F. A., Isolation of highly active photosystem II particles from a mutant of *Chlamydomonas reinhardtii, Eur. J. Biochem.,* 110, 521, 1980.
7. Thornber, J. P., Chlorophyll-proteins: reaction center and light-harvesting components of plants, *Annu. Rev. Plant Physiol.,* 26, 127, 1975.

8. Kaplan, S. and Arntzen, C. J., Photosynthetic membrane structure and function, in *Photosynthesis*, Vol. 1, Govindjee, Ed., Academic Press, New York, 1982, chap. 3.
9. Duysens, L. N. M., Transfer of Excitation Energy in Photosynthesis, Ph.D. thesis, University of Utrecht, Utrecht, The Netherlands, 1952.
10. Griffiths, M., Sistrom, W. R., Cohen-Bazire, G., and Stanier, R. Y., Function of carotenoids in photosynthesis, *Nature (London)*, 176, 1211, 1955.
11. Thornber, J. P., Maxwell, J. P., and Reinman, S., Plant chlorophyll-protein complexes: recent advances, *Photochem. Photobiol.*, 29, 1205, 1979.
12. Anderson, J. M., Barrett, J., and Thorne, S. W., Chlorophyll-protein complexes of photosynthetic eukaryotes and prokaryotes: properties and functional organization, in *Proc. 5th Int. Congr. Photosynthesis*, Vol. III, Akoyunoglou, G. A., Ed., Balaban Int. Sci. Serv., Philadelphia, 1981, 301.
13. Glazer, A. N., Comparative biochemistry of photosynthetic light-harvesting systems, *Annu. Rev. Biochem.*, 125, 1983.
14. Emerson, R. and Arnold, W., The photochemical reaction in photosynthesis, *J. Gen. Physiol.*, 16, 191, 1932.
15. Schmid, G. H. and Gaffron, H., Fluctuating photosynthetic units in higher plants and fairly constant units in algae, *Photochem. Photobiol.*, 14, 451, 1971.
16. Hill, R. and Bendall, F., Function of the two cytochrome components in chloroplasts: a working hypothesis, *Nature (London)*, 186, 136, 1960.
17. Cramer, W. A. and Crofts, A. R., Electron and proton transport, in *Photosynthesis*, Vol. 1, Govindjee, Ed., Academic Press, New York, 1982, chap. 9.
18. Anderson, J. M., Consequences of spatial separation of photosystem 1 and 2 in thylakoid membranes of higher plant chloroplasts, *FEBS Lett.*, 124, 1, 1981.
19. Anderson, J. M., The role of chlorophyll-protein complexes in the function and structure of chloroplast thylakoids, *Mol. Cell. Biochem.*, 46, 161, 1982.
20. Armond, P. A., Staehelin, L. A., and Arntzen, C. J., Spatial relationship of photosystem I, photosystem II, and the light-harvesting complex in chloroplast membranes, *J. Cell Biol.*, 73, 400, 1977.
21. Mullet, J. E., Burke, J. J., and Arntzen, C. J., Chlorophyll proteins of photosystem I, *Plant Physiol.*, 65, 814, 1980.
22. Sprague, S. G., Morschel, E., and Staehelin, L. A., Preparation of liposomes and reconstituted membranes from purified chloroplast membrane lipids, *J. Cell Biol.*, 91, 269a, 1981.
23. Mullet, J. E., Pick, U., and Arntzen, C. J., Structural analysis of the isolated chloroplast coupling factor and the N,N'-dicyclohexylcarbodiimide binding proteolipid, *Biochim. Biophys. Acta*, 642, 149, 1981.
24. Wollman, F. A., Olive, J., Bennoun, P., and Recouvreur, M., Organization of the photosystem II centers and their associated antennae in the thylakoid membranes, *J. Cell Biol.*, 87, 728, 1980.
25. Arntzen, C. J., Dynamic structural features of chloroplast lamellae, *Curr. Top. Bioenerg.*, 8, 112, 1978.
26. Andersson, B. and Akerlund, H. E., Inside-out membrane vesicles isolated from spinach thylakoids, *Biochim. Biophys. Acta*, 503, 462, 1978.
27. Boardman, N. K., Anderson, J. M. and Goodchild, D. J., Chlorophyll-protein complexes and structure of mature and developing chloroplasts, in *Curr. Top. Bioenerg.*, Vol. 8, Sanadi, D. R. and Vernon, L. P., Eds., Academic Press, New York, 1978, 35.
28. Miller, J. R. and Staehelin, L. A., Analysis of the thylakoid outer surface: coupling factor is limited to unstacked membrane regions, *J. Cell Biol.*, 68, 30, 1976.
29. Anderson, J. M., Distribution of the cytochromes of spinach chloroplasts between the appressed membranes of grana stacks and stroma-exposed thylakoid regions, *FEBS Lett.*, 138, 62, 1982.
30. Cox, R. P. and Andersson, B., Lateral and transverse organization of cytochromes in the chloroplast thylakoid membrane, *Biochim. Biophys. Res. Commun.*, 103, 1336, 1982.
31. Mörschel, E. and Staehelin, L. A., Reconstitution of cytochrome f/b₆ and ATP-synthetase from chloroplasts, in *Advances in Photosynthesis Research*, Vol. 3, Sybesma, C., Ed., W. Junk, The Hague, 1984, 251.
32. Vernon, L. P., Shaw, E. R., Ogawa, T., and Raveed, D., Structure of photosystem I and photosystem II of plant chloroplasts, *Photochem. Photobiol.*, 14, 343, 1971.
33. Malkin, R., Photosystem II activity of chloroplast fragments lacking P700, *Biochim. Biophys. Acta*, 253, 421, 1971.
34. Huzisige, H., Usiyama, H., Kikuti, T., and Azi, T., Purification and properties of the photoactive particle corresponding to photosystem II, *Plant Cell Physiol.*, 10, 441, 1969.
35. Ghirardi, M. L. and Melis, A., Localization of photosynthetic electron transport components in mesophyll and bundle sheath chloroplasts of *Zea mays*, *Arch. Biochem. Biophys.*, 224, 19, 1983.
36. Kyle, D. J., Staehelin, L. A., and Arntzen, C. J., Lateral mobility of the light harvesting complex in chloroplast membranes controls excitation energy distribution in higher plants, *Arch. Biochem. Biophys.*, 222, 527, 1983.

37. Haworth, P., Watson, J. L., and Arntzen, C. J., The detection, isolation and characterization of a light-harvesting complex which is specifically associated with photosystem I, *Biochim. Biophys. Acta,* 724, 151, 1983.
38. Stiehl, H. H. and Witt, H. T., Quantitative treatment of the function of plastoquinone in photosynthesis, *Z. Naturforsch.,* 24b, 1588, 1969.
39. Siggel, U., Renger, G., Stiehl, H. H., and Rumberg, B., Evidence for electronic and ionic interaction between electron transport chains in chloroplasts, *Biochim. Biophys. Acta,* 256, 328, 1972.
40. Velthuys, B. R., Mechanisms of electron flow in photosystem II and toward photosystem I, *Annu. Rev. Plant Physiol.,* 31, 545, 1980.
41. Sane, P. V., Goodchild, D. J., and Park, R. B., Characterization of chloroplast photosystems 1 and 2 separated by a non-detergent method, *Biochim. Biophys. Acta,* 216, 162, 1970.
42. Hauska, G. A., McCarty, R. E., Berzborn, R. J., and Racker, E., Partial resolution of the enzymes catalyzing photophosphorylation, *J. Biol. Chem.,* 246, 3524, 1971.
43. Melis, A. and Thielen, A. P. G. M., The relative absorption cross-section of photosystem I and photosystem II in chloroplasts from three types of *Nicotiana tabacum, Biochim. Biophys. Acta,* 589, 275, 1980.
44. Melis, A. and Brown, J. S., Stoichiometry of system I and system II reaction centers and of plastoquinone in different photosynthetic membranes, *Proc. Natl. Acad. Sci. U.S.A.,* 77, 4712, 1980.
45. Melis, A. and Harvey, G. W., Regulation of photosystem stoichiometry, chlorophyll *a* and chlorophyll *b* content and relation to chloroplast ultrastructure, *Biochim. Biophys. Acta,* 637, 138, 1981.
46. Anderson, M. and Melis, A., Localization of different photosystems in separate regions of chloroplast membranes, *Proc. Natl. Acad. Sci. U.S.A.,* 80, 745, 1983.
47. Melis, A., Light regulation of photosynthetic membrane structure, organization and function, *Mol. Cell. Biochem.,* in press, 1983.
48. Thielen, A. P. G. M. and Van Gorkom, H. J., Quantum efficiency and antenna size of photosystem IIα, IIβ, and I in tobacco chloroplasts, *Biochim. Biophys. Acta,* 635, 111, 1981.
49. Jursinic, P., Photosystem II/I stoichiometry by the methods of absorbance change at 325 and 705 nm, ^{14}C-atrazine binding, and oxygen flash yields, in *Advances in Photosynthesis Research,* Vol. 3, Sybesma, C., Ed., W. Junk, The Hague, 1984, 485.
50. Whitmarsh, J. and Ort, D. R., Quantitative determination of the electron transport complexes in the thylakoid membrane of spinach and several other plant species, in *Advances in Photosynthesis Research,* Vol. 3, Sybesma, C., Ed., W. Junk, The Hague, 1984, 231.
51. Emerson, R. and Lewis, C. M., The dependence of the quantum yield of *Chlorella* photosynthesis on wavelength of light, *Am. J. Bot.,* 30, 165, 1943.
52. Mullet, J. E. and Arntzen, C. J., Regulation of grana stacking in a model membrane system: mediation by a purified light-harvesting pigment-protein complex from chloroplasts, *Biochim. Biophys. Acta,* 635, 23, 1980.
53. Butler, W. L., Energy distribution in the photochemical apparatus of photosynthesis, *Annu. Rev. Plant Physiol.,* 29, 345, 1978.
54. Anderson, J. M., The significance of grana stacking in chlorophyll *b*-containing chloroplasts, *Photochem. Photobiophys.,* 3, 225, 1982.
55. Bonaventura, C. and Myers, J., Fluorescence and oxygen evolution from *Chlorella pyrenoidosa, Biochim. Biophys. Acta,* 189, 366, 1969.
56. Murata, N., Control of excitation transfer in photosynthesis. II. Magnesium ion-dependent distribution of excitation energy between two pigment systems in spinach chloroplasts, *Biochim. Biophys. Acta,* 189, 171, 1969.
57. Myers, J., Enhancement studies in photosynthesis, *Annu. Rev. Plant Physiol.,* 2, 282, 1971.
58. Haworth, P., Kyle, D. J., Horton, P., and Arntzen, C. J., Chloroplast membrane protein phosphorylation, *Photochem. Photobiol.,* 36, 743, 1982.
59. Staehelin, L. A. and Arntzen, C. J., Regulation of chloroplast membrane function: protein phosphorylation changes in the spatial organization of membrane components, *J. Cell Biol.,* 97, 1327, 1983.
60. Bennett, J., Phosphorylation of chloroplast membrane polypeptides, *Nature (London),* 269, 344, 1977.
61. Bennett, J., Chloroplast phosphoproteins. Phosphorylation of polypeptides of the light-harvesting chlorophyll protein complex, *Eur. J. Biochem.,* 99, 133, 1979.
62. Bennett, J., Chloroplast phosphoproteins. Evidence for a thylakoid-bound phosphoprotein phosphatase, *Eur. J. Biochem.,* 104, 84, 1980.
63. Horton, P. and Black, M. T., Activation of adenosine 5'-triphosphate-induced quenching of chlorophyll fluorescence by reduced plastoquinone, *FEBS Lett.,* 119, 141, 1980.
64. Allen, J. F., Bennett, J., Steinbeck, K. E., and Arntzen, C. J., Chloroplast protein phosphorylation couples redox state to distribution of excitation energy between photosystems, *Nature (London),* 291, 25, 1981.
65. Bennett, J., Steinbeck, K., and Arntzen, C. J., Chloroplast phosphoproteins: regulation of excitation energy transfer by phosphorylation of thylakoid membrane polypeptides, *Proc. Natl. Acad. Sci. U.S.A.,* 77, 5253, 1980.

66. Williams, W. P., The two photosystems and their interactions, in *Topics in Photosynthesis*, Vol. 2, Barber, J., Ed., Elsevier, Amsterdam, 1977, chap. 3.
67. Kok, B., Efficiency of photosynthesis, in *Horizons of Bioenergetics*, San Pietro, A. and Gest, H., Eds., Academic Press, New York, 1972, 153.
68. Bolton, J. R., Solar energy conversion efficiency in photosynthesis — or — why two photosystems?, *Proc. 4th Int. Congr. Photosynthesis*, Hall, D. O., Coombs, J., and Goodwin, T. W., Eds., Biochemical Society, London, 1978, 621.
69. Bolton, J. R. and Hall, D. O., Photochemical conversion and storage of solar energy, *Annu. Rev. Energy*, 4, 353, 1979.
70. Coombs, J., Prospects of improving photosynthesis for increased crop productivity, in *Advances in Photosynthesis Research*, Vol. 4, Sybesma, C., Ed., W. Junk, The Hague, 1984, 85.
71. Ku, S. B. and Edwards, G. E., Quantum requirement of photosynthetic bacteria and plants, in *Handbook of Biosolar Resources, Vol. 1, Basic Principles*, Mitsui, A. and Black, C. C., Eds., CRC Press, Boca Raton, Fla., 1982, 37.
72. Loomis, R. S. and Williams, W. A., Maximum crop productivity: an estimate, *Crop Sci.*, 3, 67, 1963.
73. Zelitch, I., *Photosynthesis, Photorespiration and Plant Productivity*, Academic Press, New York, 1971, 279.
74. Gaastra, P., Photosynthesis of crop plants as influenced by light, carbon dioxide, temperature and stomatal diffusion resistance, *Meded. Landbouwhogesch. Wageningen*, 59, 1, 1959.
75. Black, C. C., Photosynthetic carbon fixation in relation to net CO_2 uptake, *Annu. Rev. Plant Physiol.*, 24, 252, 1973.
76. Bassham, J. A. and Buchanan, B. B., Carbon dioxide fixation pathways in plants and bacteria, in *Photosynthesis, Vol. II, Development, Carbon Metabolism, and Plant Productivity*, Govindjee, Ed., Academic Press, New York, 1982, chap. 6.
77. Witt, H. T., Coupling of quanta, electrons, fields ions, and phosphorylation in the functional membranes of photosynthesis, *Q. Rev. Biophys.*, 4, 365, 1971.
78. Haehnel, W., Electron transport between plastoquinone and chlorophyll *a* in chloroplasts, *Biochim. Biophys. Acta*, 305, 618, 1973.
79. Björkman, O., Boardman, N. K., Anderson, J. M., Thorne, S. W., Goodchild, D. J., and Pyliotis, N. A., Effect of light intensity during growth of *Atriplex patula* on the capacity of photosynthetic reactions, chloroplast components and structure, *Carnegie Inst. Washington Yearb.*, 71, 115, 1972.
80. Boardman, N. K., Björkman, O., Anderson, J. M., Goodchild, D. J., and Thorne, S. W., Photosynthetic adaptation of higher plants to light intensity: relationship between chloroplast structure, composition of the photosystems and photosynthetic rates, in *Proc. 3rd Int. Congr. Photosynthesis*, Avron, M., Ed., Elsevier, Amsterdam, 1974, 1809.
81. Boardman, N. K., Comparative photosynthesis of sun and shade plants, *Annu. Rev. Plant Physiol.*, 28, 355, 1977.
82. Patterson, D. J., Light and temperature adaptation, in *Predicting Photosynthesis for Ecosystem Models*, Vol. 1., Hesketh, J. D., Ed., CRC Press, Boca Raton, Fla., 1980, chap. 8.
83. Björkman, O., Responses to different quantum flux densities, *Encycl. Plant Physiol. New Ser.*, 12A, 57, 1981.
84. Berry, J. A. and Downton, W. J. S., Environmental regulation of photosynthesis, in *Photosynthesis, Vol. 2, Development, Carbon Metabolism and Plant Productivity*, Govindjee, Ed., Academic Press, New York, 1982, chap. 9.
85. Lichtenthaler, H. K., Burgstahler, R., Buschmann, C., Meier, D., Prenzel, U., and Schonthal, A., Effect of high light and high light stress on composition, function and structure of the photosynthetic apparatus, in *Effects of Stress on Photosynthesis*, Marcelle, R., Clijsters, H., and Van Poucke, M., Eds., Martinus Nijhoff/W. Junk, The Hague, 1983, 353.
86. Björkman, O., Ludlow, M. M., and Morrow, P. A., Photosynthetic performance of two rainforest species in their native habitat and analysis of their gas exchange, *Carnegie Inst. Washington Yearb.*, 71, 94, 1972.
87. Goodchild, D. J., Björkman, O., and Pyliotis, N. A., Chloroplast ultrastructure, leaf anatomy, and content of chlorophyll and soluble protein in rainforest species, *Carnegie Inst. Washington Yearb.*, 71, 102, 1972.
88. Anderson, J. M., Goodchild, D. J., and Boardman, N. K., Composition of the photosystems and chloroplast structure in extreme shade plants, *Biochim. Biophys. Acta*, 325, 573, 1973.
89. Björkman, O., Comparative studies on photosynthesis in higher plants, *Photophysiology*, 8, 1, 1973.
90. Alberte, R. S., McClure, P. R., and Thornber, J. P., Photosynthesis in trees, organization of chlorophyll and photosynthetic unit size in isolated gymnosperm chloroplasts, *Plant Physiol.*, 58, 341, 1976.
91. Malkin, S. and Fork, D. C., A comparative study of the photosynthetic unit sizes in sun and shade plants, *Carnegie Inst. Washington Yearb.*, 79, 187, 1980.
92. Fork, D. C. and Govindjee, Chlorophyll *a* fluorescence transients of leaves from sun and shade plants, *Naturwissenshaften*, 67, 510, 1980.

93. Sheridov, R. P., Adaptation to quantum flux by the Emerson photosynthetic unit, *Plant Physiol.*, 50, 355, 1972.

94. Patterson, D. T., Duke, S. O., and Hoagland, R. E., Effects of irradiance during growth on adaptive photosynthetic characteristics of velvet leaf and cotton, *Plant Physiol.*, 61, 402, 1978.

95. Leong, T.-Y. and Anderson, J. M., Changes in composition and function of thylakoid membranes as a result of photosynthetic adaptation of chloroplasts from pea plants grown under different light conditions, *Biochim. Biophys. Acta*, 723, 391, 1983.

96. Grahl, H. and Wild, A., Studies on the content of P_{700} and cytochromes of *Sinapis alba* during growth under two different light intensities, in *Environmental and Biological Control of Photosynthesis*, Marcelle, R., Ed., W. Junk, The Hague, 1975, 107.

97. Berzborn, R. J., Muller, D., Roos, P., and Anderson, B., Significance of different quantitative determinations of photosynthetic ATP-synthetase CF_1 for heterogeneous CF_1 distribution and grana formation, in *Proc. 5th Int. Congr. Photosynthesis, Vol. III,* Akoyunoglou, G., Ed., Balaban Int. Sci. Serv., Philadelphia, 1981, 107.

98. Björkman, O., Environmental and biological control of photosynthesis: inaugural address, in *Environmental and Biological Control of Photosynthesis*, Marcelle, R., Ed., W. Junk, The Hague, 1975, 1.

99. Levine, R. P., The analysis of photosynthesis using mutant strains of algae and higher plants, *Annu. Rev. Plant Physiol.*, 20, 523, 1969.

100. Kirk, J. T. O. and Tilney-Bassett, R. A. E., *The Plastids*, Freeman, London, 1967.

101. Boardman, N. K. and Highkin, H. R., Studies on a barley mutant lacking chlorophyll *b*. I. Photochemical activity of isolated chloroplasts, *Biochim. Biophys. Acta*, 126, 189, 1966.

102. Schmid, G. H. and Gaffron, H., Light metabolism and chloroplast structure in chlorophyll deficient tobacco mutants, *J. Gen. Physiol.*, 50, 563, 1967.

103. Highkin, H. R., Boardman, N. K., and Goodchild, D. J., Photosynthetic studies on a pea-deficient mutant in chlorophyll, *Plant Physiol.*, 44, 1310, 1969.

104. Keck, R. W., Dilley, R. A., Allen, C. F., and Biggs, S., Chlorophyll composition and structure differences in a soybean mutant, *Plant Physiol.*, 46, 692, 1970.

105. Keck, R. W., Dilley, R. A., and Ke, B., Photochemical characteristics in a soybean mutant, *Plant Physiol.*, 46, 699, 1970.

106. Bazzaz, M. B., Govindjee, and Paolillo, D. J., Biochemical, spectral, and structural study of olive necrotic 8147 mutant of *Zea mays* L., *Z. Pflanzenphysiol.*, 72, 181, 1974.

107. Benedict, C. R. and Kohel, R. J., Characteristics of a virescent cotton mutant, *Plant Physiol.*, 43, 1611, 1968.

108. Okave, K., Schmid, G. H., and Straub, J., Characterization and high efficiency photosynthesis of an aurea mutant of tobacco, *Plant Physiol.*, 60, 150, 1977.

109. Nielsen, N. C., Smillie, R. M., Henningsen, K. W., and Von Wettstein, D., Composition and function of thylakoid membranes from grana rich and grana deficient chloroplast mutants of barley, *Plant Physiol.*, 63, 174, 1979.

110. Hopkins, W. G., German, J. B., and Hayden, P. B., A light-sensitive mutant in maize. II. Photosynthetic properties, *Z. Pflanzenphysiol.*, 100, 15, 1980.

111. Benedict, C. R., McCree, K. J., and Kohel, R. J., High photosynthetic rate of a chlorophyll mutant of cotton, *Plant Physiol.*, 49, 968, 1972.

112. Alberte, R. S., Hesketh, J. D., Hofstra, G., Thornber, J. P., Naylor, A. W., Bernard, R. L., Brim, C., Endrizzi, J., and Kohel, R. J., Structure and activity of the photosynthetic apparatus in temperature-sensitive mutants of higher plants, *Proc. Natl. Acad. Sci. U.S.A.*, 71, 2414, 1974.

113. Schmid, G. H., Origin and properties of mutant plants: yellow tobacco, *Methods Enzymol.*, 23, 171, 1971.

114. Arntzen, C. J. and Briantais, J.-M., Chloroplast structure and function, in *Bioenergetics of Photosynthesis*, Govindjee, Ed., Academic Press, 1975, chap. 2.

115. Eskins, K., Delmastro, D., and Harris, L., A comparison of pigment-protein complexes among normal, chlorophyll-deficient and senescent soybean genotypes, *Plant Physiol.*, 73, 51, 1983.

116. Highkin, H. R. and Frenkel, A. W., Studies of growth and metabolism of a barley mutant lacking chlorophyll *b*, *Plant Physiol.*, 37, 814, 1962.

117. Schmid, G. H. and Gaffron, H., Quantum requirement for photosynthesis in chlorophyll-deficient plants with unusual lamellar structures, *J. Gen. Physiol.*, 50, 2131, 1967.

118. Benedict, C. R. and Kohel, R. J., Photosynthetic rate of a virescent cotton mutant lacking chloroplast grana, *Plant Physiol.*, 45, 519, 1970.

119. Alberti, R. S., Hesketh, J. D., and Kirby, J. S., Comparison of photosynthetic activity and lamellar characteristics of virescent and normal green peanut leaves, *Z. Pflanzenphysiol.*, 77, 152, 1976.

120. Prioul, J.-L., Limiting factors in photosynthesis from the chloroplast to the plant canopy, in *Trends in Photobiology*, Helen, C., Charlier, M., Monteray-Garestier, T., and Laustriat, G., Eds., Plenum Press, New York, 1982, 633.

121. Gifford, R. M. and Jenkins, C. L. D., Prospects of applying knowledge of photosynthesis toward crop production, in *Photosynthesis, Vol. II, Development, Carbon Metabolism, and Plant Productivity,* Govindjee, Ed., Academic Press, 1982, chap. 12.

122. Moss, D. N., Studies on increasing photosynthesis in crop plants, in *CO_2 Metabolism and Plant Productivity,* Burris, R. H. and Black, C. C., Eds., University Park Press, Baltimore, 1976, 31.

123. Hardy, R. W. F., Havelka, V. D., and Quebedeaux, B., Increasing crop productivity: the problem, strategies, approach, and selected rate-limitations related to photosynthesis, in *Proc. 4th Int. Congr. Photosynthesis,* Hall, D. O., Coombs, J., and Goodwin, T. W., Eds., The Biochemical Society, London, 1978, 695.

124. Ogren, W. L. and Chollet, R., Photorespiration, in *Photosynthesis, Vol. II, Development, Carbon Metabolism, and Plant Productivity,* Govindjee, Ed., Academic Press, 1982, chap. 7.

125. Miziorko, H. M. and Lorimer, G. H., Ribulose-1,5-bisphosphate carboxylase-oxygenase, *Annu. Rev. Biochem.,* 52, 507, 1983.

126. Wittwer, S. H. and Robb, W. M., Carbon dioxide enrichment of greenhouse atmospheres for food crop production, *Econ. Bot.,* 18, 34, 1964.

127. Ogren, W. L., Increasing carbon fixation by crop plants, in *Proc. 4th Int. Congr. Photosynthesis,* Hall, D. O., Coombs, I., and Goodwin, T. W., Eds., The Biochemical Society, London, 1978, 721.

128. Terry, N., Limiting factors in photosynthesis. IV. Iron stress-mediated changes in light-harvesting and electron transport capacity and its effects on photosynthesis *in vivo, Plant Physiol.,* 71, 855, 1983.

129. Terry, N., Control of photosynthetic rate: influence of light-harvesting and electron transport capacity in different environments, in *Advances in Photosynthesis Research,* Vol. 4, Sybesma, C., Ed., W. Junk, The Hague, 1984, 233.

130. Terry, N., Limiting factors in photosynthesis. I. Use of iron stress to control photochemical capacity *in vivo, Plant Physiol.,* 65, 114, 1980.

131. Taylor, S. E., Terry, N., and Huston, R. P., Limiting factors in photosynthesis. III. Effects of iron nutrition on the activities of three regulatory enzymes of photosynthetic carbon metabolism, *Plant Physiol.,* 70, 1541, 1982.

132. Terry, N., The use of mineral nutrient stress in the study of limiting factors in photosynthesis, in *Photosynthesis and Plant Development,* Marcelle, R., Slijsters, H., and Van Poucke, Ed., W. Junk, The Hague, 1979, 151.

133. Platt-Aloia, K. A., Thomson, W. W., and Terry, N., Changes in plastid ultrastructure during iron nutrition-mediated chloroplast development, *Protoplasma,* 114, 85, 1983.

134. Spiller, S. and Terry, N., Limiting factors in photosynthesis. II. Iron stress diminishes photochemical capacity by reducing the number of photosynthetic units, *Plant Physiol.,* 65, 121, 1980.

135. Nishio, J. N. and Terry, N., Iron nutrition-mediated chloroplast development, *Plant Physiol.,* 71, 688, 1983.

136. Willstätter, R. and Stoll, A., *Untersuchungen uber die Assimilation der Kohlensaure,* Springer-Verlag, Berlin, 1918, 135.

137. Emerson, R., The relation between maximum rate of photosynthesis and concentration of chlorophyll, *J. Gen. Physiol.,* 12, 609, 1929.

138. Sesták, Z., Limitations for finding a linear relationship between chlorophyll content and photosynthetic activity, *Biol. Plant.,* 8, 336, 1966.

139. Buttery, B. R. and Buzzell, R. I., The relationship between chlorophyll content and rate of photosynthesis in soybeans, *Can. J. Plant Sci.,* 57, 1, 1977.

140. Gabrielson, E. K., Effects of different chlorophyll concentrations on photosynthesis in foliage leaves, *Physiol. Plant.,* 1, 5, 1948.

141. Hesketh, J. D., Limitations to photosynthesis responsible for differences among species, *Crop Sci.,* 3, 493, 1963.

142. Ferguson, H., Eslick, R. F., and Aase, J. K., Canopy temperatures of barley are influenced by morphological characteristics, *Agron. J.,* 65, 425, 1973.

143. McCashin, B. G. and Canvin, D. T., Photosynthetic and photorespiratory characteristics of mutants of *Hordeum vulgare* L., *Plant Physiol.,* 64, 354, 1979.

144. Lilley, R. M. and Walker, D. A., Carbon dioxide assimilation by leaves, isolated chloroplasts, and ribulose bisphosphate carboxylase from spinach, *Plant Physiol.,* 55, 1087, 1975.

145. Collatz, G. J., The interactions between photosynthesis and ribulose-P_2 concentration-effects of light, CO_2 and O_2, *Carnegie Inst. Washington Yearb.,* 77, 248, 1978.

146. VonCaemmerer, S. and Farquhar, G. D., Some relationships between the biochemistry of photosynthesis and the gas exchange of leaves, *Planta,* 153, 376, 1981.

147. Gifford, R. M. and Evans, L. T., Photosynthesis, carbon partitioning, and yield, *Annu. Rev. Plant Physiol.,* 32, 485, 1981.

148. Elmore, C. D., The paradox of no correlation between leaf photosynthetic rates and crop yields, in *Predicting Photosynthesis for Ecosystem Models,* Vol. 2, Hesketh, J. D. and Jones, J. W., Eds., CRC Press, Boca Raton, Fla., 1980, chap. 9.

149. Crosbie, T. M., Mock, J. J., and Pearce, R. B., Relationship among CO_2-exchange rate and plant traits in Iowa stiff stalk synthetic maize population, *Crop Sci.*, 15, 476, 1975.
150. Nelson, D. J., Asay, K. H., and Horst, G. L., Relationship of leaf photosynthesis to forage yield of tall fescue, *Crop Sci.*, 18, 405, 1978.
151. Wilhelm, W. W. and Nelson, C. J., Irradiance response of tall fescue genotypes with contrasting levels of photosynthesis and yield, *Crop. Sci.*, 18, 405, 1978.
152. Murthy, K. K. and Singh, M., Photosynthesis, chlorophyll content and ribulose diphosphate carboxylase in relation to yield in wheat genotypes, *J. Agric. Sci.*, 93, 7, 1979.
153. Grafius, J. E. and Barnard, J., Leaf canopy as related to yield in barley, *Agron. J.*, 68, 398, 1976.
154. Delaney, R. H. and Dobrenz, A. K., Yield of alfalfa as related to carbon exchange, *Agron. J.*, 66, 498, 1974.
155. Curtis, P. E., Ogren, W. L., and Hageman, R. H., Varietal effects in soybean photosynthesis and photorespiration, *Crop Sci.*, 9, 323, 1969.
156. Irvine, J. E., Relations of photosynthetic rates and leaf and canopy characters to sugarcane yield, *Crop Sci.*, 15, 671, 1975.
157. Hart, R. H., Pearce, R. B., Chatterton, N. J., Carlson, G. E., Barnes, D. K., and Hanson, C. H., Alfalfa yield, specific leaf weight, CO_2 exchange rate and morphology, *Crop Sci.*, 18, 649, 1978.
158. Evans, L. T., The physiological basis of crop yield, in *Crop Physiology*, Evans, L. T., Ed., Cambridge University Press, Cambridge, England, 1975, chap. 11.
159. Pendleton, J. W., Egli, D. B., and Peters, D. B., Response of *Zea mays* L. to a "light rich" field environment, *Agron. J.*, 59, 395, 1967.
160. Jong, S. K., Brewbaker, J. L., and Lee, C. H., Effects of solar radiation in the performance of maize in 41 successive monthly plantings in Hawaii, *Crop Sci.*, 22, 13, 1982.
161. Hesketh, J. D. and Moss, D. N., Variation in the response of photosynthesis to light, *Crop Sci.*, 3, 107, 1963.
162. Johnston, T. J., Pendleton, J. W., Peters, D. B., and Hicks, D. R., Influence of supplemental light on apparent photosynthesis, yield, and yield components of soybeans (*Glycine max* L.), *Crop Sci.*, 9, 577, 1969.
163. Johnston, T. J. and Pendleton, J. W., Contribution of leaves at different canopy levels to seed production of upright and lodged soybeans (*Glycine max* L.), *Crop Sci.*, 8, 291, 1968.
164. Christy, A. L. and Porter, C. A., Canopy photosynthesis and yield in soybean, in *Photosynthesis, Vol. II, Development, Carbon Metabolism, and Plant Productivity*, Govindjee, Ed., Academic Press, New York, 1982, chap. 14.
165. Wells, R., Schulze, L. L., Ashley, D. A., Boerma, H. R., and Brown, R. H., Cultivar differences in canopy apparent photosynthesis and their relationship to seed yield in soybeans, *Crop Sci.*, 22, 886, 1982.
166. Hirschberg, J. and McIntosh, L., Molecular basis of herbicide resistance in *Amaranthus hybridus*, *Science*, 222, 1346, 1983.

Chapter 5

RIBULOSE BISPHOSPHATE CARBOXYLASE: PROPERTIES AND REGULATION

James T. Bahr

TABLE OF CONTENTS

I. INTRODUCTION

Ribulose-1,5-bisphosphate carboxylase-oxygenase may be regarded as the starting point for two major metabolic pathways in green plant tissues. Photosynthetic CO_2 assimilation in a certain sense begins with the carboxylation of ribulose-1,5-bisphosphate, catalyzed by this protein, to form two molecules of 3-phosphoglycerate. The role of the photochemical and bioenergetic metabolism phases of photosynthesis may be envisioned as converting the resultant 3-phosphoglycerate into higher energy sugar phosphates for use by other metabolic pathways (triose phosphates), and as providing the thermodynamic driving force for the carboxylation reaction (by regenerating ribulose-1,5-bisphosphate). The photorespiratory pathway also begins at a reaction catalyzed by this protein, the oxygenation of ribulose-1,5-bisphosphate to yield one molecule of 3-phosphoglycerate and one molecule of 2-phosphoglycolate. After dephosphorylation of 2-phosphoglycolate, the glycolic acid leaves the chloroplast and is metabolized with loss of CO_2 via enzymes located in various cellular compartments.

Ribulose-1,5-bisphosphate carboxylase-oxygenase is one of the most abundant proteins in the biosphere.[1] The amount present in leaf tissue varies, of course, with nutrient levels and plant species, but is of the order of 6 mg per mg chlorophyll.[2] Using a chlorophyll content of 3 mg/dm^2 leaf area, and a leaf area index of 3 to 4, one can calculate that the average well-kept suburban lawn has about 2 to 3 kg of ribulose-1,5-bisphosphate carboxylase-oxygenase per acre.

The quantity of this protein present in leaf tissue accounts for its other noncatalytic functions in plant metabolism and agriculture. As the major protein component in leaf tissue, it becomes the major protein component in forage crops and for grazing livestock. The amino acid composition of the protein is generally well-balanced,[3] with amounts of essential amino acids which equal or exceed FAO recommendations. In recognition of this amino acid composition, several groups in recent years have conducted research into isolation and purification methods for the protein for use as a human food supplement (in the form of Rubisco Crackers?). Two U.S. patents on such methodology have been issued.[4]

Ribulose-1,5-bisphosphate carboxylase-oxygenase also serves as a major form of storage of organic nitrogen in the plant. Some 50% of the nitrogen which eventually is used in synthesis of grain proteins in soybeans is first stored in the leaf predominately as the carboxylase-oxygenase.[5] During the later stages of bean growth in soybeans, leaf tissue senesces and amino acids are transferred from the soluble leaf proteins to the developing beans.

It has been suggested recently that the carboxylase-oxygenase also serves as a metabolite buffer in the chloroplast stroma.[6] It is known that many of the components of the reductive pentose phosphate cycle in the stroma bind to the carboxylase at concentrations which are physiologically realistic.[7] The carboxylase itself is present in the stroma at a concentration of 3- to 5-mM subunits and thus has a large potential for binding major percentages of sugar phosphates, adenosine phosphates, and NADPH. In the light in vivo, the levels of ribulose-1,5-bisphosphate present are, however, also several millimolar, and, as the binding constant for ribulose-1,5-bisphosphate is higher than for most other metabolites,[8] it will occupy the vast majority of available sites. In the dark in vivo, ribulose-1,5-bisphosphate concentration is much lower and binding of stromal phosphates to the carboxylase must be considered potentially significant. A second situation where binding of metabolites may be significant is in the isolated chloroplast, where in the light the levels of ribulose-1,5-bisphosphate are normally very much lower than in intact leaves,[9] providing many more potential binding sites for other metabolites.

The most significant role for ribulose-1,5-bisphosphate carboxylase-oxygenase in metabolite buffering would appear to be with respect to NADPH/NADP⁺ and ATP/

ADP/Pi. Based on older studies of activation and inhibition by these compounds,[10] the carboxylase binds one form of each pair (NADPH and ATP) much more tightly than the other form. This would alter the true redox state and phosphorylation potential of the stroma from the values calculated from assays of total pyridine and adenine nucleotides.

II. BIOCHEMISTRY OF RIBULOSE 1,5-BISPHOSPHATE CARBOXYLASE

A. Isolation Methods

There are essentially two techniques for isolation of ribulose-1,5-bisphosphate carboxylase. The first technique consists of ammonium sulfate precipitation of the protein after elimination of chlorophyll containing materials at lower ammonium sulfate concentrations, followed by separation from lower molecular weight proteins by sucrose density gradient centrifugation or one or more cycles of gel permeation chromatography. This sort of preparation is free of contaminants by most physical assays, but this "success" is primarily the result of the large amount of carboxylase protein initially present. The preparation may not be free of contaminating enzyme activities (e.g., see Reference 11). Elimination of these other enzymes may require further purification steps.

The second approach is via crystallization. The carboxylase may be crystallized directly from relatively crude leaf extracts either by dialysis against low ionic strength media (*Nicotiana* species)[12] or by vapor diffusion against polyethylene glycol (for more species).[13] The degree of purity achieved by these methods has not been well characterized.

Success in isolation of active carboxylase is dependent on protection against damage by plant phenolics during initial tissue homogenization. The specific activity of the crystalline tobacco enzyme appears to be dependent on the growing conditions of the tobacco plants used.[14] The worst case for phenolic damage is probably glanded cotton, where, with the usual methodology, it is not possible to prepare any carboxylase, active or inactive. The phenolics result in the precipitation of all soluble protein. Inclusion of borate in the extraction procedure[15] permits not only protein recovery, but active carboxylase recovery with an initial specific activity of about 1.0 μmol CO_2 (mg protein $\times$ min)$^{-1}$.

B. Assay Methods

The observed activity of ribulose-1,5-bisphosphate carboxylase is always influenced by the history of the enzyme sample being used, with the conditions during 1 to 10 min prior to initiation of the assay being most critical. For routine assays, the pretreatment of the enzyme with Mg^{2+} and CO_2 at concentrations and pH values sufficient to give "full activation" is the most reliable.[16] If the Mg^{2+} and CO_2 concentrations or the pH of the assay itself is then different, or if incompletely activated enzyme is used, the possibility of a nonconstant rate of reaction must be considered, as the degree of activation changes in response to the assay environment. Inhibition of activity at longer assay times by ribulose-1,5-bisphosphate or impurities therein is also commonly observed.[17]

Ribulose-1,5-bisphosphate carboxylase-oxygenase is also subject to partial inactivation upon storage at 4°C for a period of hours or more.[18] This inactivation is reversible, with treatment for 2 hr at 37 to 40°C or 10 min at 55°C being sufficient to reactivate.[19]

The most widely used assay methods for ribulose-1,5-bisphosphate carboxylase-oxygenase are measurement of $H^{14}CO_3^-$ incorporation into acid-stable products for the carboxylase activity, and polarographic O_2 consumption for the oxygenase activity.[16]

Other available methods include spectrophotometric coupled enzyme assays for 3-phosphoglycerate production[20-21] and colorometric assay for glyoxylate derived from 2-phosphoglycolate via phosphoglycolate phosphatase and glycolate oxidase.[22]

Two recently described assays utilizing labeled ribulose-1,5-bisphosphate were developed to permit convenient simultaneous assay of both carboxylase and oxygenase activities.[23-24] Simultaneous assay of both activities is necessary for reliable studies of relative rates of the two competing reactions. For some values of O_2 and CO_2 concentrations, the simultaneous assay can be conducted via $H^{14}CO_3^-$ and polarographic O_2 uptake measurement but this is not always convenient.

The first method[23] utilizes [1-^{14}C, 5-^{3}H] ribulose-1,5-bisphosphate and measures ^{3}H and ^{14}C radioactivity in the product, 3-phosphoglycerate after separation from starting material, and 2-phosphoglycolate by ion exchange chromatography. The ^{14}C 3-phosphoglycerate arises only from carboxylation, the ^{3}H 3-phosphoglycerate arises from both reactions. Enrichment of 3-phosphoglycerate in ^{3}H vs. ^{14}C is proportional to oxygenase activity.

The second method[24] utilizes [1-^{3}H] ribulose 1,5-bisphosphate and $^{14}CO_2$. The rate of incorporation of ^{14}C into acid stable 3-phosphoglycerate measures carboxylation. The rate of production of ^{3}H-phosphoglycolate measures oxygenation. The ^{3}H-phosphoglycolate is measured as ^{3}H-glycolate after treatment with phosphoglycolate phosphatase and ion exchange chromatography to separate ^{3}H-glycolate from ^{3}H-phosphoglycerate and ^{3}H-ribulose-1,5-bisphosphate. The reliability of the assay is dependent on the specificity of the phosphoglycolate phosphatase used. Dephosphorylation of ^{3}H-phosphoglycerate must be avoided.

Under some conditions, the observed activity of ribulose-1,5-bisphosphate carboxylase-oxygenase can be dependent on the addition of carbonic anhydrase. Such a dependence was the basis on which the identity of the substrate as CO_2 rather than HCO_3^- was first established.[25] At slightly alkaline pH values, the bulk of CO_2 in solution exists as HCO_3^-, and the pool of CO_2 might be exhausted by the carboxylation reaction rather rapidly. The rate of hydration of HCO_3^- is usually rapid enough to replenish the CO_2 before significant departure from the initial, equilibrium value can occur. In cases of especially high enzyme concentration and reaction rates, and low CO_2/HCO_3^- concentrations, the intrinsic rate of HCO_3^- hydration may not be fast enough and carbonic anhydrase is required for maximal rates. Two recent reports are based on these considerations.[26-27] The suggestion to include carbonic anhydrase in routine assays involving low [$\leqslant$ the $KmCO_2$)] CO_2 concentrations seems worthwhile.

C. Physical Properties

Ribulose-1,5-bisphosphate carboxylase-oxygenase is a protein of about 550,000 kdalton molecular weight, containing eight of each of two sizes of subunits.[3,28] The subunit molecular weights are in the range of 12,000 to 15,000 kdalton and 52,000 to 56,000 kdalton. The exception to this is the enzyme from some photosynthetic bacteria where the protein is a dimer of large subunits only. Other than progress in the gross geometry[29] made possible by crystallization of the protein (see later), little new information on classical physical properties has developed since the reviews cited above.

The carboxylase contains almost no heavy metal ions when isolated and purified under stringent conditions. Crystals from eight different species[13] contained not more than 0.40 g atom of copper and 0.20 g atom of iron per eight large subunits. Similar low values have been noted for preparation by other methods also.[31-32] The crystalline spinach enzyme assayed for oxygenase activity in a medium containing less than 0.2 g atom heavy metal per eight subunits had a specific oxygenase activity of 0.08 μmol O_2/mg protein/min at 21% O_2 after a 1 min exposure to deactivating conditions. This is about what would be expected for a good preparation and suggests that heavy metal ions (Cu or Fe) are not co-factors of the oxygenase activity.

The small subunit displays a diversity of charged species when subjected to isoelectric focusing.[33] Amino acid sequence differences between the two apparent polypeptides in *Nicotiana tabacum* have been demonstrated to account for the charge difference.[34] In *Pisum sativum,* the amino acid sequence has been measured by direct sequencing[35] and by cDNA sequencing.[36] The *Pisum* small subunit also displays two charge species on isoelectric focusing experiments, but the amino acid sequence has only one site of variation, the N terminus. Both methionine and glutamine were found at this site, in a relationship with valine at position 2/3 suggesting that the absence of methionine was a translational or posttranslational event. Cyclization of the exposed N-terminal glutamine could account for the two charge isomers. The isoelectric focusing data and sequence results for the spinach small subunit are similar to those for *Pisum,*[37] with 89 out of 123 residues being identical.

The functional role of the small subunit remains unclear. Although aggregates of large subunits have rather low specific activities,[38] this may be a consequence of the subunit-dissociating treatments rather than evidence for a controlling role of the small subunit. In the *Rhodospirillum rubrum* carboxylase, which is a dimer of large subunits only, the pattern of activation by CO_2 and Mg^{2+} and the effects of certain sugar phosphates on activation is qualitatively similar to that in higher plant enzymes.[39] Thus, earlier suggestions for a "regulatory" role for the small subunit cannot be related to any known regulatory phenomenon.

The amino acid sequence of the large subunit has been determined for spinach[40] and maize[41] via sequencing of the DNA coding for its synthesis. Comparison of the two sequences indicates 16% divergence between the two species, but a high degree of conservation in the active site regions (AA 166-178, 321-340, and 450-462). The large subunit from many species demonstrates a threefold charge diversity in isoelectric focusing experiments.[42-44] This is not necessarily evidence for three distinct nuclear genes, however, as the changes may have arisen after transcription or during isolation. The carboxylase from wheat, isolated by differential precipitation with ammonium sulfate followed by Sepharose 6B and DEAE-cellulose chromatography, when subjected to isoelectric focusing after dissociation of subunits and derivatization with iodoacetamide, gave several bands corresponding to large subunits and one small subunit band.[45] When the enzyme preparation was carried out without the use of ammonium sulfate precipitation, either by preparative gel electrophoresis or by immunoprecipitation of subunits, the isoelectric focusing experiments yielded a single predominant large subunit band. This single band could be split into multiple bands by increasing the molar ratio of iodoacetamide to large subunit used during protein derivatization prior to the focusing experiment. At ratios of 200 to 300, the typical three polypeptide pattern seen in most other studies was generated. In these studies, complex multiple large subunit bands were observed for all preparations based on ammonium sulfate precipitation, regardless of iodoacetamide concentration used. Further, no effects on the single small subunit isoelectric focusing band were obtained by variation of iodoacetamide concentration. Careful consideration of conditions used and interpretations made in prior isoelectric focusing experiments is required.

D. Catalytic Properties

The catalytic process of ribulose-1,5-bisphosphate carboxylase-oxygenase is characterized by a Km (ribulose-1,5-bisphosphate) of about 30 μM, a Km (CO_2) of from 10 μM to 70 μM, and a Km (O_2) of 500 to 700 μM.[16,28,46] The maximal velocity for carboxylation is about 2 μmol (mg protein × min)$^{-1}$, but frequently lower, presumably due to irreversible inactivation during isolation. Evidence using carboxyarabinitol bisphosphate has suggested the true maximal velocity is 2.8 μmol (mg protein × min)$^{-1}$ for the spinach enzyme.[47] Maximal velocities of the oxygenase reaction are one third to one quarter that of the carboxylase reaction.[48,49] The literature on the mechanism of the reaction from a bio-organic chemical view has been reviewed rather recently[46] and is not considered in detail here. The effects of protein modification reagents and active-site directed inhibitors have implicated lysine,[50-55] cysteine,[55] histidine,[56,57] arginine,[58,59] tyrosine,[50] and methionine[61] in the catalytic mechanism.

In the past few years, several lines of investigation have been concerned with the possibility of "improving" the kinetic properties of ribulose-1,5-bisphosphate carboxylase. From an agricultural productivity viewpoint, an "improvement" in the carboxylase would occur if the relative ratio of carboxylase to oxygenase activities could be increased at prevailing CO_2 and O_2 concentrations. A variety of factors have been claimed to alter carboxylase to oxygenase ratios; most all of these have been examined by subsequent researches and only two have been confirmed. The oxygenase to carboxylase ratio can be modified by the choice of metal ion used during activation and assay, as the metal ion alters the kinetic constants for O_2 and CO_2. Genetic factors may also alter the ratio by determining the apparent Km [CO_2] for catalysis.

It has been known for a long time that metal ions other than Mg^{2+} were capable of serving as activators/cofactors for the carboxylase.[62] The bivalent cation, shown to be required for formation of the active enzyme species,[67,68] is probably also involved in catalysis. Binding of carboxyribitol bisphosphate,[69] a transition state analog, altered the interaction of Mn^{2+} and CO_2 measured by NMR methods. Manganese (II) has shown a reproducible ability to alter carboxylase-oxygenase ratios in favor of increased oxygenase.[49,63-66] When utilized as the activator metal ion, Mn^{2+} stimulated oxygenase activity to levels greater than observed with Mg^{2+}, while giving only minimal carboxylase activity. The transition metals CO^{2+} and Ni^{2+} also gave a similar response.[64] The differential response of the two activities of the enzyme to metal ions is a manifestation of changes in both maximal velocity and gas substrate affinities. In soybean carboxylase-oxygenase,[66] the maximal velocities with Mg^{2+}/Mn^{2+} were 1.70/0.29 for carboxylation and 0.61/0.29 for oxygenation (in μmol [mg protein × min]$^{-1}$). The Km (CO_2) was 2.5 mM HCO_3 with Mg^{2+} and 0.85 mM with Mn^{2+}; the Km (O_2) was 37.0% with Mg^{2+} and 1.7% with Mn^{2+}. In a subsequent study,[49] these values were confirmed for both soybean and *R. rubrum* enzyme and determined for Fe^{2+}, C^{2+}, and Ni^{2+}.

The modification of catalytic regulation properties by metal ions other than Mg^{2+} is, however, entirely in the wrong direction for achieving enhanced photosynthesis or decreased photorespiration. In the data set of maximal velocity and substrate Km values generated in these studies, there exists a moderate correlation of maximal velocity with substrate affinity.[49] Reduction of Km (CO_2) reduces carboxylation maximal velocity. Increasing Km (O_2) increases oxygenase maximal velocity. In both cases, it is the Mg^{2+} data which lies most outside the general relationship for the other metals. This result suggests that K_{cat} contributes importantly to the substrate Km values. A study of the effect of Mg^{2+} and Mn^{2+} on apparent pK values characterizing maximal velocity - pH curves and maximal velocity over Km (CO_2) vs. pH curves supports the concept that the metal ion effect is largely on K_{cat} rather than on gas substrate binding per se.[70] The apparent pKa at 8.1 in Vmax/Km (CO_2) vs. pH curves was unaffected by replacement of Mn^{2+} for Mg^{2+}.

Changes in oxygenase to carboxylase activities and in substrate affinities have been reported among and within species. Four diploid cultivars of rye were reported[71] to possess reduced CO_2 affinity while O_2 affinity and maximal velocities were normal (comparison is with four tetraploid cultivars). Although this result was indirectly confirmed by studies of the CO_2 affinity of isolated rye protoplasts,[72] a subsequent study in two laboratories[73,74] of isolated rye carboxylase-oxygenase found normal, i.e., tetraploid-type, kinetic constants for diploid rye cultivars. The role of ploidy in influencing ribulose-1,5-bisphosphate carboxylase kinetics must therefore be regarded as unproven.

The Km (CO_2) of ribulose-1,5-bisphosphate carboxylase from various species of plants has been determined by several groups. Determination of Km (CO_2) values for catalysis, free of effects of CO_2 concentration on activation, requires careful attention to details of the assay. A defined state of activation (preferably 100%), very careful pH control, and short assay times (to reduce deactivation at low CO_2 concentrations) are all critical. All of the groups cited below appear to be fully aware of these considerations. The Km (CO_2) values in freshly lysed chloroplast extracts, after activation, of six C-3 species varied from 7.8 μM to 12.4 μM.[75] This variation was not observed for the 3 species in common with a subsequent study of 16 plant species.[76] In this study, Km (CO_2) values of nine C-3 species ranged from 10.2 μM to 14.0 μM. The only species with a Km (CO_2) widely different from these was maize, at 27 μM. An overall mean standard error of 5% was reported.

Surveys of Km (CO_2) for 60 grass species[77] and 54 assorted species,[78] including C-3, C-4, and aquatic species, have been made. The C-3 species had Km values from 12 to 26 μM, and the C-4 species had Km values from 28 to 63 μM. Aquatic plants had Km (CO_2) values from 30 to 70 μM. The standard deviations are reported to be 2 to 3 μM, so that even within photosynthetic metabolism types, statistically significant variations exist.

In light of the effects on CO_2 and O_2 affinities by metal ions and their linkage to maximal velocity, a decrease in Km (CO_2) alone is an inadequate measure of potential for enhanced photosynthesis or agronomic advantage for a species or variety. Both Km (CO_2) and Km (O_2) and their associated maximal velocities must be measured. For a selection of C-3, C-4, aquatic plants, and photosynthetic bacteria, this has been done.[48] The Km (CO_2) values varied from 9 to 16 μM for five C-3 plants and were 16 μM and 34 μM for two C-4 plants. In each case, the concomitant variation in Km (O_2) or in maximal velocities was such that no important change in overall specificity for carboxylation over oxygenation resulted (maximum effect, 6%).

In summary, the existing data does not support the hope that the selectivity of ribulose-1,5-bisphosphate carboxylase-oxygenase can be improved. An improvement in CO_2 affinity such that the enzyme operates at a greater percentage of its maximal velocity, even if oxygenase to carboxylase ratios remain constant, should still be advantageous.

III. ACTIVATION AND BIOSYNTHESIS OF RIBULOSE-1,5-BISPHOSPHATE CARBOXYLASE

A. Activation

Ribulose-1,5-bisphosphate carboxylase-oxygenase is subject to regulation of catalytic activity by the degree of formation of an enzyme-CO_2-Mg^{2+} complex,[67] via interaction of CO_2 with a lysine residue on the large subunit to produce a covalent carbamate.[79] The metal ion appears to bind to the carbamate and shift the equilibrium towards carbamate formation. The equilibrium may be further shifted in favor of the active enzyme species by a large series of inorganic ions, organic phosphates, and related compounds.[7,80,81] The existing evidence is consistent with the proposal that these known effectors bind at the catalytic site itself. Positive effectors presumably bind more strongly to the catalytic site of the active enzyme than of the inactive enzyme, and also dissociate rapidly when the assay per se is begun with addition of ribulose-1,5-bisphosphate. Positive effectors added simultaneously with ribulose-1,5-bisphosphate are either inactive or competitive inhibitors.[81] Negative effectors presumably bind more tightly to the inactive enzyme or fail to vacate the catalytic site in favor of ribulose-1,5-bisphosphate. Since effectors bind at the catalytic site, an enzyme molecule with an effector cannot be catalytically competent,[7] and in the steady state these effectors can have no role in increasing carboxylase activity.

It should be noted that although all known activating effectors of the enzyme appear to bind at the catalytic site of the active form, there is no *a priori* reason why there cannot be other molecules capable of preferential binding at other sites of the active enzyme.

Activation and deactivation of ribulose-1,5-bisphosphate carboxylase are physiological phenomena in isolated chloroplasts and protoplasts. The degree of activation may be assayed by hypotonic or surfactant lysis of the membranes in a carboxylase-oxygenase assay medium.[82,83] In leaves, the existing activation state may be preserved during tissue extraction by working at $2°C$.[84]

As the rate of photosynthesis depends not only on the kinetic properties of the carboxylase, but also the concentration of active enzyme species present, the degree of activation in vivo is a possible limiting factor for photosynthesis. The percent activation in illuminated isolated chloroplasts and protoplasts at physiological CO_2 levels[82,83,85] is high (>50%) but not necessarily complete. Activation can be decreased by incubation at very low CO_2 concentrations, and the CO_2 concentration required for half-maximal activation is dependent on light and photosynthetic electron transport.[82,85]

In whole leaves of spinach and wheat,[84] grown in an environmental chamber, the percent activation of ribulose-1,5-bisphosphate carboxylase was from 20 to 63%, depending on leaf age and species. Values between 50 and 70% have been noted for environmental chamber grown soybeans.[100] Under limiting light intensity in 7- to 8-day wheat seedlings, activation of carboxylase varied with light intensity in exact proportion to the variation in photosynthetic rate,[8] suggesting a role for carboxylase activation in controlling photosynthesis. Activation did not reach zero in darkness, however.

Considerable additional work, including studies in field grown crops, is required to reach an understanding of the regulation of carboxylase-oxygenase activation in vivo, and to determine whether activation is partial and rate-limiting, or complete, under the high light intensities of a field crop.

Several observations indicate that regulation of activation of ribulose-1,5-bisphosphate carboxylase in vivo is rather different from that in vitro. First, the level of activation observed at physiological CO_2 and Mg^{2+} is much higher than expected from in vitro studies. Second, ribulose-1,5-bisphosphate fails to inhibit enzyme activation in vivo,[86] while it is a negative effector in vitro. Third, a mutant of *Arabidopsis thaliana* has been identified[87] in which light and CO_2 do not control in vivo activation while in vitro activation is normal.

B. Biosynthesis

The details of the biosynthesis of large and small subunits of ribulose-1,5-bisphosphate carboxylase-oxygenase are beyond the concerns of this article. The question of the regulation of the amount of carboxylase-oxygenase synthesized is not. If the enzyme operates in vivo at suboptimal levels because its CO_2 affinity is limiting, and if the chloroplast photosystems are capable of supplying ribulose-1,5-bisphosphate at an adequate rate, then more carboxylase per chloroplast (or reaction center) should result in more photosynthesis. It is clear that the photosystems can supply ribulose-1,5-bisphosphate at a higher rate, for they do so at elevated CO_2 levels.

Amounts of carboxylase-oxygenase on a leaf area or percent of total protein basis can vary[88,89] and are influenced by light intensity and nitrogen nutrition. Apparently, genetic factors can also control carboxylase levels.

In tall fescue, carboxylase content increased with ploidy level from a mean of 43 to 55% of total protein,[90] while maintaining a constant specific activity. Variations in carboxylase within a ploidy level were also observed. The increases in carboxylase were associated with an increase in chlorophyll such that carboxylase to chlorophyll ratios tended to remain constant. This result suggests that the effect of ploidy increase was primarily to increase chloroplast number per leaf area. If leaf area also decreased, no net advantage would result. The increase in carboxylase as a percent of total protein, however, suggests that more of the resources of the leaf are committed to photosynthesis at high ploidy in tall fescue. Associated with this was a mean 47% increase in photosynthesis on a leaf dry weight basis.

The relationships of ploidy level in wheat to carboxylase-oxygenase levels on a leaf area and chloroplast basis do not confirm the above increased commitment to photosynthesis.[91,92] Three ploidy levels (2×, 4×, and 6×) were examined. Carboxylase per cell increased in direct proportion to the increase in DNA per cell; carboxylase to DNA remained constant. Chloroplast number per cell also increased with ploidy, but not quite as rapidly as carboxylase per cell. Carboxylase per chloroplast therefore increased. The changes in carboxylase per plastid were not, however, statistically significant. In contrast to the tall fescue results, carboxylase as a percent of total protein did not change. Rather than alter the cell's commitment of resources to photosynthetic apparatus biosynthesis, the change in ploidy level has primarily altered cell size, and thus leaf thickness and chloroplasts per cell.

A similar result was obtained in alfalfa, comparing 2×, 4×, and 8× ploidy levels.[92,93] Ribulose-1,5-bisphosphate carboxylase per cell or protoplast doubled for the 2× vs. 4× comparison, as did many other parameters including protoplast photosynthesis on a per cell basis. For the 4× vs. 8× comparison, less than a full doubling was observed. No evidence was obtained for preferential biosynthesis of carboxylase-oxygenase with changed ploidy level. Ploidy level altered instead the pattern of cellular packaging.

IV. OPPORTUNITIES FOR PLANT BREEDING AND GENETIC ENGINEERING

Ribulose-1,5-bisphosphate carboxylase-oxygenase in C-3 plants is clearly a yield-limiting factor of major importance in agriculture. Photosynthesis in C-3 plants increased when ambient CO_2 was increased at high light intensities. Increased photosynthesis achieved by CO_2 enrichment led, in soybeans under field conditions, to an increased yield of 98%.[94] Somewhat smaller yield increases were obtained for peas and dry beans.

For nonlegumes and for determinate C-3 plants, the response to enhanced photosynthesis may be different. In C-3 plants where N_2 fixation driven by photosynthetic products is not possible, achievement of higher yields from increased photosynthesis may also require increased fertilizer inputs to support the greater biomass accumulation. In some C-3 crops, the response to increased photosynthesis by CO_2 enrichment does not lead to increased yield, but to the same yield with an earlier maturity.[95] In this type of crop (determinate growth pattern), the improvement of ribulose-1,5-bisphosphate carboxylase would result in new opportunities for improved agronomic practices; e.g., double-cropping, rather than in yield increases.

It is possible to estimate the magnitude of increased yield expected from increased photosynthesis for a crop such as soybeans. Early in the growth of a field crop, increased photosynthesis would be expected to produce a very large effect through a "compound interest" relationship. The increased photosynthate availability should increase vegetative growth and the rate of leaf area expansion. Thus, not only will photosynthesis per leaf area increase, but the percent of sunlight utilized for photosynthesis increases as well. As the crop approaches a closed canopy, which should be earlier in the season at higher photosynthetic rates, the effect of increased photosynthesis should decline. For a closed canopy crop, a 10% increase in photosynthesis will result in simply a 10% increase in dry weight accumulation. In soybeans, the bulk (75%) of the total dry weight accumulated during the season is accumulated after canopy closure,[96,97] and the duration of photosynthesis under closed canopy conditions, increased by an earlier canopy closure date at higher photosynthesis, will be at least as important quantitatively as the increased rate of photosynthesis by itself.

A two-week earlier canopy closure coupled with a 10% increase in closed-canopy photosynthesis should lead to a 25 to 30% increase in seasonal dry weight accumulation. If the harvest index (total reproductive growth divided by total growth) remained constant, this would give a 25 to 30% yield increase. There is evidence,[98,99] however, that harvest index will decline somewhat with increased photosynthesis, leading to a yield increase in the 15 to 25% range.

It has been the general experience of plant breeders that selection for increased photosynthesis does not lead to increased yield. We suggest that most examples of varieties with increased photosynthesis are, in fact, varieties with lower chlorophyll levels or a thicker leaf structure, rather than varieties with increased photosynthetic efficiency. This body of work does not rule out the idea of increased photosynthesis leading to increased yields.

It is our conclusion that small increases in the CO_2 affinity of ribulose-1,5-bisphosphate carboxylase, 10 to 20%, could have quantitatively significant effects on crop productivity. Similarly, small increases in the extent of enzyme activation or in the level of total carboxylase-oxygenase in the chloroplast could also be quantitatively important for crop production.

REFERENCES

1. Ellis, R. J., The most abundant protein in the world, *Trends Biochem. Sci.*, 4, 241, 1979.
2. Zelitch, I., *Photosynthesis, Photorespiration and Plant Productivity*, Academic Press, New York, 1971, xiv.
3. Kawashima, N. and Wildman, S. G., Fraction I protein, *Annu. Rev. Plant Physiol.*, 21, 325, 1970.
4. U.S. Patents No. 3823128 and 4340676.
5. Wittenbach, V. A., Breakdown of ribulose bisphosphate carboxylase and change in proteolytic activity during dark-induced senescence of wheat seedlings, *Plant Physiol.*, 62, 604, 1978.
6. Ashton, A. R., A role for ribulose-1,5-bisphosphate carboxylase as a metabolite buffer, *FEBS Lett.*, 145, 1, 1982.
7. Badger, M. R. and Lorimer, G. H., Interaction of sugar phosphates with the catalytic site of ribulose-1,5-bisphosphate carboxylase, *Biochemistry*, 20, 2219, 1981.
8. Perchorowitz, J. T., Raynes, D. A., and Jensen, R. G., Light limitation of photosynthesis and activation of ribulose bisphosphate carboxylase in wheat seedlings, *Proc. Natl. Acad. Sci. U.S.A.*, 78, 2985, 1981.
9. Sicher, R. D. and Jensen, R. G., Photosynthesis and ribulose 1,5-bisphosphate levels in intact chloroplasts, *Plant Physiol.*, 64, 880, 1979.
10. Chu, D. K. and Bassham, J. A., Activation of ribulose 1,5-bisphosphate carboxylase by nicotinamide, adenine dinucleotide phosphate and other chloroplast metabolites, *Plant Physiol.*, 54, 556, 1974.
11. Miziorko, H. M., Nowak, T., and Mildvan, A. S., Spinach leaf phosphoenolpyruvate carboxylase: purification, properties, and kinetic studies, *Arch. Biochem. Biophys.*, 163, 378, 1974.
12. Chan, P. H., Sakano, K., Singh, S., and Wildman, S. G., Crystalline fraction I protein: preparation in large yield, *Science*, 176, 1145, 1972.
13. Johal, S., Bourque, D. P., Smith, W. W., Suh, S. W., and Eisenberg, D., Crystallization and characterization of ribulose 1,5-bisphosphate, *J. Biol. Chem.*, 255, 8873, 1980.
14. Bahr, J. T., Johal, S., Capel, M., and Bourque, D. P., High specific activity ribulose 1,5-bisphosphate carboxylase-oxygenase from nicotiana tabacum, *Photosyn. Res.*, 2, 235, 1981.
15. Quayle, T. J., Katterman, F. R., and Jensen, R. G., Isolation of active ribulose 1,5-bisphosphate carboxylase from glanded cotton *(G. hirsutum)*, *Physiol. Plant.*, 50, 233, 1980.
16. Lorimer, G. H., Badger, M. R., and Andrews, T. J., D-ribulose-1,5-bisphosphate carboxylase-oxygenase. Improved methods for the activation and assay for catalytic activities, *Anal. Biochem.*, 78, 66, 1977.

17. Paech, C., Pierce, J., McCurry, S. D., and Tolbert, N. E., Inhibition of ribulose-1,5-bisphosphate carboxylase/oxygenase by ribulose-1,5-bisphosphate epimerization and degradation products, *Biochem. Biophys. Res. Commun.*, 83, 1084, 1978.
18. Chollet, R. and Anderson, L. L., Conformational changes associated with the reversible cold inactivation of ribulose-1,5-bisphosphate carboxylase-oxygenase, *Biochim. Biophys. Acta*, 482, 228, 1977.
19. Kung, S. D., Chollet, R., and Marsho, T., Crystallization and assay procedures of tobacco ribulose-1,5-bisphosphate carboxylase-oxygenase, *Methods Enzymol.*, 69, 326, 1980.
20. Rice, S. C. and Pon, N. G., Direct spectrophotometric observation of ribulose-1,5-bisphosphate carboxylase activity, *Anal. Biochem.*, 87, 39, 1978.
21. Lilley, R. M. and Walker, D. A., An improved spectrophotometric assay for ribulose-1,5-bisphosphate carboxylase from spinach, *Biochim. Biophys. Acta*, 358, 226, 1974.
22. Laing, W. A., Ogren, W. L., and Hageman, R. H., Regulation of soybean net photosynthetic CO_2 fixation by the interaction for CO_2, O_2, and ribulose-1,5-diphosphate carboxylase, *Plant Physiol.*, 54, 678, 1974.
23. Kent, S. S. and Young, J. D., Simultaneous kinetic analysis of ribulose-1,5-bisphosphate carboxylase/oxygenase, *Plant Physiol.*, 65, 465, 1980.
24. Jordan, D. B. and Ogren, W. L., A sensitive assay, procedure for simultaneous determination of ribulose-1,5-bisphosphate carboxylase and oxygenase activities, *Plant Physiol.*, 67, 237, 1981.
25. Cooper, T. G., Filmer, D., Wishnick, M., and Lane, M. D., The active species of CO_2 utilized by ribulose diphosphate carboxylase, *J. Biol. Chem.*, 244, 1081, 1969.
26. Shiraiwa, Y. and Miyachi, S., Enhancement of ribulose 1,5-bisphosphate carboxylation reaction by carbonic anhydrase, *FEBS Lett.*, 106, 243, 1979.
27. Bird, I. F., Cornelius, M. J., and Keys, A. J., Effect of carbonic anhydrase on the activity of ribulose 1,5-bisphosphate carboxylase, *J. Exp. Bot.*, 31, 365, 1980.
28. Jensen, R. G. and Bahr, J. T., Ribulose 1,5-bisphosphate carboxylase-oxygenase, *Annu. Rev. Plant Physiol.*, 28, 379, 1977.
29. Baker, T. S., Eisenberg, D., and Eiserling, F., Ribulose 1,5-bisphosphate carboxylase: a two-layered, square-shaped molecule of symmetry 422, *Science*, 196, 293, 1977.
30. Chollet, R., Anderson, L. L., and Hovsepian, L. C., The absence of tightly bound copper, iron, and flavin nucleotide in crystalline ribulose 1,5-bisphosphate carboxylase-oxygenase from tobacco, *Biophys. Res. Commun.*, 64, 97, 1975.
31. Lorimer, G. H., Andrews, T. J., and Tolbert, N. E., Ribulose diphosphate oxygenase. II. Further proof of reaction products and mechanism of action, *Biochemistry*, 12, 18, 1973.
32. Hall, N. P., McCurry, S. D., and Tolbert, N. E., Storage and maintaining activity of ribulose bisphosphate carboxylase/oxygenase, *Plant Physiol.*, 67, 1220, 1981.
33. Kung, S. D., Sakano, K., and Wildman, S. G., Multiple peptide composition of the large and small subunits of *Nicotiana tabacum* fraction I protein ascertained by fingerprinting and electrofocusing, *Biochim. Biophys. Acta*, 365, 138, 1974.
34. Strobaek, S., Gibbons, G. C., Haslett, B. G., Boulter, D., and Wildman, S. G., On the nature of the polymorphism of the small subunit of ribulose 1,5-bisphosphate carboxylase in the amphidiploid *Nicotiana tabacum*, *Carlsberg Res. Commun.*, 41, 335, 1976.
35. Takruri, I. A. H., Boulter, D., and Ellis, R. J., Amino acid sequence of the small subunit of ribulose 1,5-bisphosphate carboxylase of *Pisum sativum*, *Phytochemistry*, 20, 413, 1981.
36. Bedbrook, J. R., Smith, S. M., and Ellis, R. J., Molecular cloning and sequencing of cDNA encoding the precursor to the small subunit of chloroplast ribulose 1,5-bisphosphate carboxylase, *Nature (London)*, 287, 692, 1980.
37. Martin, P. G., Amino acid sequence of the small subunit of ribulose 1,5-bisphosphate carboxylase from spinach, *Aust. J. Plant Physiol.*, 6, 401, 1979.
38. Takabe, T. and Akzawa, T., Catalytic role of subunit A in ribulose 1,5-diphosphate carboxylase from chromatium strain D., *Arch. Biochem. Biophys.*, 157, 303, 1973.
39. Whitman, W. B., Martin, M. N., and Tabita, F. R., Activation and regulation of ribulose bisphosphate carboxylase-oxygenase in the absence of small subunits, *J. Biol. Chem.*, 254, 10184, 1979.
40. Zurawski, G., Perrot, B., Bottomley, W., and Whitfield, P. R., The structure of the gene for the large subunit of ribulose 1,5-bisphosphate carboxylase from spinach chloroplasts, *Nucleic Acids Res.*, 9, 3251, 1981.
41. McIntosh, L., Poulsen, C., and Bogorad, L., Chloroplast gene sequence for the large subunit of ribulose bisphosphate carboxylase of maize, *Nature (London)*, 288, 556, 1980.
42. Chen, K., Kung, S. D., Gray, J. C., and Wildman, S. G., Subunit polypeptide composition of fraction I protein from various plant species, *Plant Sci. Lett.*, 7, 429, 1976.
43. Uchimiya, H., Chen, K., and Wildman, S. G., Evolution of fraction I protein in the genus lycopersicon, *Biochem. Genet.*, 17, 333, 1979.
44. Gatenby, A. A. and Cocking, E. C., Fraction I protein and the origin of the european potato, *Plant Sci. Lett.*, 12, 177, 1978.

45. O'Connell, P. B. H. and Brady, C. J., Multiple forms of the large subunit of wheat ribulose bisphosphate carboxylase generated by excell iodoacetamide, *Biochim. Biophys. Acta,* 670, 355, 1981.

46. Lorimer, G. H., The carboxylation and oxygenation of ribulose 1,5-bisphosphate: the primary events in photosynthesis and photorespiration, *Annu. Rev. Plant Physiol.,* 32, 349, 1981.

47. Hall, N. P., Pierce, J., and Tolbert, N. E., Formation of carboxyarabinitol bisphosphate complex with ribulose bisphosphate carboxylase/oxygenase and theoretical specific activity of the enzyme, *Arch. Biochem. Biophys.,* 212, 115, 1981.

48. Jordan, D. B. and Ogren, W. L., Species variation in the specificity of ribulose bisphosphate carboxylase/oxygenase, *Nature (London),* 291, 513, 1981.

49. Christeller, J. T., The effects of bivalent cations on ribulose bisphosphate carboxylase/oxygenase, *Biochem. J.,* 193, 839, 1981.

50. Paech, C. and Tolbert, N. E., Active site studies of ribulose 1,5-bisphosphate carboxylase/oxygenase with pyridoxal 5'-phosphate, *J. Biol. Chem.,* 253, 7864, 1978.

51. Stringer, C. D. and Hartman, F. C., Sequences of two active site peptides from spinach ribulose bisphosphate carboxylase/oxygenase, *Biochem. Biophys. Res. Commun.,* 80, 1043, 1978.

52. Whitman, W. B. and Tabita, F. R., Modification of *Rhodospirillum rubrum* ribulose bisphosphate carboxylase with pyridoxal phosphate. I. Identification of a lysyl residue of the active site, *Biochemistry,* 17, 1283, 1978.

53. Herndon, C. S., Norton, I. L., and Hartman, F. C., Reexamination of the binding site for pyridoxal 5'-phosphate in ribulose bisphosphate carboxylase/oxygenase from *Rhodospirillum rubrum, Biochemistry,* 21, 1380, 1982.

54. Chollet, R. and Anderson, L. L., Cyanate modification of essential lysyl residues in the catalytic subunit of tobacco ribulose bisphosphate carboxylase, *Biochim. Biophys. Acta,* 525, 455, 1978.

55. Schloss, J. V., Stringer, C. D., and Hartman, F. C., Identification of essential lysyl and cysteinyl residues in spinach ribulose bisphosphate carboxylase/oxygenase modified by the affinity label N-bromoacetylethanolamine phosphate, *J. Biol. Chem.,* 253, 5707, 1978.

56. Bhagwat, A. S. and Ramakrishna, J., Essential histidine residues of ribulose bisphosphate carboxylase indicated by reaction with diethylpyrocarbonate and rose bengal, *Biochem. Biophys. Acta,* 662, 181, 1981.

57. Saluja, A. K. and McFadden, B. A., Modification of active site histidine in ribulose bisphosphate carboxylase/oxygenase, *Biochemistry,* 21, 89, 1982.

58. Schloss, J. V., Norton, I. L., Stringer, C. D., and Hartman, F. C., Inactivation of ribulose bisphosphate carboxylase by modification of arginyl residues with phenylglyoxal, *Biochemistry,* 17, 5626, 1978.

59. Chollet, R., Inactivation of crystalline tobacco ribulose bisphosphate carboxylase by modification of arginine residues with 2,3-butanedione and phenylglyoxal, *Biochim. Biophys. Acta,* 658, 177, 1981.

60. Robison, P. D. and Tabita, F. R., Modification of ribulose bisphosphate carboxylase from *Rhodospirillum rubrum* with tetranitromethane, *Biochem. Biophys. Res. Commun.,* 88, 85, 1979.

61. Christeller, J. T. and Hartman, F. C., Inactivation of *Rhodospirillum rubrum* ribulose bisphosphate carboxylase/oxygenase by the affinity label 2-N-chloramino-2-deoxypentitol 1,5-bisphosphate, *FEBS Lett.,* 142, 162, 1982.

62. Weissbach, A., Horecker, B. L., and Hurwitz, J., The enzymic formation of phosphoglyceric acid from ribulose diphosphate and carbon dioxide, *J. Biol. Chem.,* 218, 795, 1956.

63. Wildner, G. F. and Henkel, J., Differential reactivation of ribulose 1,5-bisphosphate oxygenase with low carboxylase activity by Mn^{2+}, *FEBS Lett.,* 91, 99, 1978.

64. Wildner, G. F. and Henkel, J., The effect of divalent, metal ions on the activity of Mg^{2+}, depleted ribulose-1,5-bisphosphate oxygenase, *Planta,* 146, 223, 1979.

65. Robison, P. D., Martin, M. N., and Tabita, F. R., Differential effects of metal ions on *Rhodospirillum rubrum* ribulose bisphosphate carboxylase/oxygenase and stoichiometric incorporation of HCO_3 — into a cobalt (III) — enzyme complex, *Biochemistry,* 18, 4453, 1979.

66. Christeller, J. T. and Laing, W. A., Effects of manganese ions and magnesium ions on the activity of soya-bean ribulose bisphosphate carboxylase/oxygenase, *Biochem. J.,* 183, 747, 1979.

67. Lorimer, G. H., Badger, M. R., and Andrews, T. J., The activation of ribulose-1,5-bisphosphate carboxylase by carbon dioxide and magnesium ions. Equilibria, kinetics, a suggested mechanism, and physiological implications, *Biochemistry,* 15, 529, 1976.

68. O'Leary, M. H., Jaworski, R. J., and Hartman, F. C., ^{13}C nuclear magnetic resonance of the CO_2 activation of ribulose bisphosphate carboxylase from *Rhodospirillum rubrum, Proc. Natl. Acad. Sci. U.S.A.,* 76, 673, 1979.

69. Miziorko, H. M. and Mildvan, A. S., Electron paramagnetic resonance, ^{1}H, and ^{13}C nuclear magnetic resonance studies of the interaction of manganese and bicarbonate with ribulose 1,5-diphosphate carboxylase, *J. Biol. Chem.,* 249, 2743, 1974.

70. Christeller, J. T., Effects of divalent cations on the activity of ribulose bisphosphate carboxylase: interactions with pH and with D_2O as solvent, *Arch. Biochem. Biophys.*, 217, 485, 1982.
71. Garrett, M. K., Control of photorespiration at RuBP carboxylase/oxygenase level in ryegrass cultivars, *Nature (London)*, 274, 913, 1978.
72. Rathnam, C. K. M. and Chollet, R., Photosynthetic and photorespiratory carbon metabolism in mesophyll protoplasts and chloroplasts isolated from isogenic diploid and tetraploid cultivars of ryegrass (*Lolium perenne* L.), *Plant Physiol.*, 65, 489, 1980.
73. McNeil, P. H., Foyer, C. H., Walker, D. A., Bird, I. F., Cornelius, M. J., and Keyes, A. J., Similarity of ribulose-1,5-bisphosphate carboxylases of isogenic diploid and tetraploid ryegrass (*Lolium perenne* L.) cultivars, *Plant Physiol.*, 67, 530, 1981.
74. Rejda, J. M., Johal, S., and Chollet, R., Enzymic and physicochemical characteristics of ribulose-1,5-bisphosphate carboxylase/oxygenase from diploid and tetraploid cultivars of ryegrass (*Lolium perenne* L.), *Arch. Biochem. Biophys.*, 210, 617, 624, 1981.
75. Delaney, M. E. and Walker, D. A., Comparison of the kinetic properties of ribulose bisphosphate carboxylase in chloroplast extracts of spinach, sunflower, and four other reductive pentose phosphate-pathway species, *Biochem. J.*, 171, 477, 1978.
76. Bird, I. F., Cornelius, M. J., and Keyes, A. J., Affinity of RuBP carboxylases for carbon dioxide and inhibition of the enzymes by oxygen, *J. Exp. Bot.*, 33, 1004, 1982.
77. Yeoh, H. H., Badger, M. R., and Watson, L., Variations in $Km(CO_2)$ of ribulose-1,5-bisphosphate carboxylase among grasses, *Plant Physiol.*, 66, 1110, 1980.
78. Yeoh, H. H., Badger, M. R., and Watson, L., Variations in kinetic properties of ribulose-1,5-bisphosphate carboxylases among plants, *Plant Physiol.*, 67, 1151, 1981.
79. Lorimer, G. H. and Miziorko, H. M., Carbamate formation on the amino group of a lysyl residue as the basis for the activation of ribulose bisphosphate carboxylase by CO_2 and Mg^{2+}, *Biochemistry*, 19, 5321, 1980.
80. Hatch, A. L. and Jensen, R. G., Regulation of ribulose-1,5-bisphosphate carboxylase from tobacco: changes in pH response and affinity for CO_2 and Mg^{2+} induced by chloroplast intermediates, *Arch. Biochem. Biophys.*, 295, 587, 1980.
81. McCurry, S. D., Pierce, J., Tolbert, N. E., and Orme-Johnson, W. H., On the mechanism of effector-mediated activation of ribulose bisphosphate carboxylase/oxygenase, *J. Biol. Chem.*, 256, 6623, 1981.
82. Bahr, J. T. and Jensen, R. G., Activation of ribulose bisphosphate carboxylase in intact chloroplasts by CO_2 and light, *Arch. Biochem. Biophys.*, 185, 39, 1978.
83. Robison, S. P., McNeil, P. H., and Walker, D. A., Ribulose bisphosphate carboxylase: lack of dark inactivation of the enzyme in experiments with protoplasts, *FEBS Lett.*, 97, 296, 1979.
84. Perchorowitz, J. T., Raynes, D. A., and Jensen, R. G., Measurement and preservation of the *in vivo* activation of ribulose-1,5-bisphosphate carboxylase in leaf extracts, *Plant Physiol.*, 69, 1165, 1982.
85. Sicher, R. C., Reversible light-activation of ribulose bisphosphate carboxylase/oxygenase in isolated barley protoplasts and chloroplasts, *Plant Physiol.*, 70, 336, 1982.
86. Sicher, R. C., Hatch, A. L., Stumpf, D. K., and Jensen, R. G., Ribulose-1,5-bisphosphate and activation of the carboxylase in the chloroplast, *Plant Physiol.*, 68, 252, 1981.
87. Somerville, C. R., Portis, A. R., and Ogren, W. L., A mutant of *Arabidopsis thaliana* which lacks activation of RuBP carboxylase *in vivo*, *Plant Physiol.*, 70, 381, 387, 1982.
88. Wildman, S. G., Aspects of fraction I protein evaluation, *Arch. Biochem. Biophys.*, 196, 598, 1979.
89. Ku, M. S. B., Schmitt, M. R., and Edwards, G. E., Quantitative determination of RuBP carboxylase-oxygenase protein in leaves of several C-3 and C-4 plants, *J. Exp. Bot.*, 30, 89, 1978.
90. Joseph, M. C., Randall, D. D., and Nelson, C. J., Photosynthesis in polyploid tall fescue. II. Photosynthesis and ribulose-1,5-bisphosphate carboxylase of polyploid tall fescue, *Plant Physiol.*, 68, 894, 1981.
91. Dean, C. and Leech, R. M., Genome expression during normal leaf development. II. Direct correlation between ribulose bisphosphate carboxylase content and nuclear ploidy in polyploid series of wheat, *Plant Physiol.*, 70, 1605, 1982.
92. Meyers, S. P., Nichols, S. L., Baer, G. R., Molin, W. T., and Schraeder, L. E., Ploidy effects in isogenic populations of alfalfa. I. Ribulose-1,5-bisphosphate carboxylase, soluble protein, chlorophyll, and DNA in leaves, *Plant Physiol.*, 70, 1704, 1982.
93. Molin, W. T., Meyers, S. P., Baer, G. R., and Schraeder, L. E., Ploidy effects in isogenic populations of alfalfa. II. Photosynthesis, chloroplast number, ribulose-1,5-bisphosphate carboxylase, chlorophyll, and DNA in protoplasts, *Plant Physiol.*, 70, 1710, 1982.
94. Hardy, R. W. F., Havelka, U. D., and Quebedeaux, B., The opportunity for and significance of alteration of ribulose-1,5-bisphosphate carboxylase activities in crop production, in *Photosynthetic Carbon Assimilation*, Siegelman, H. W. and Hind, G., Eds., Plenum Press, New York, 1978, 165.
95. Mauney, J. F., Fry, K. E., and Guinn, G., Relationship of photosynthetic rate to growth and fruiting of cotton, soybean, sorghum, and sunflower, *Crop Sci.*, 18, 259, 1978.

96. Hanway, J. J. and Weber, C. R., Dry matter accumulation in eight soybean (*Glycine max* L. Merrill) varieties, *Agron. J.*, 63, 227, 1971.
97. Hanway, J. J., Interrelated developmental and biochemical processes in the growth of soybean plants, in *World Soybean Research*, Hill, L. D., Ed., Interstate Printers and Publ., Danville, Ill., 1976, 5.
98. Quebedeaux, B. and Hardy, R. W. F., Reproductive growth and dry matter production of *Glycine max* (L.) Merr. in response to oxygen concentration, *Plant Physiol.*, 55, 102, 1975.
99. Cooper, R. L. and Brun, W. A., Response of soybeans to a carbon dioxide-enriched atmosphere, *Crop Sci.*, 7, 455, 1967.
100. Bahr, J. T., unpublished observations.

Chapter 6

EFFICIENCY OF CARBON ASSIMILATION

Clanton C. Black, Jr.

TABLE OF CONTENTS

I. INTRODUCTION

Photosynthesis is our primary process for assimilating net quantities of CO_2 for utilization by all biological organisms and, within the last two decades, our understanding of photosynthesis has expanded remarkably. Our expanded understanding currently includes such features of photosynthesis as its diversity: in carbon biochemistry; in responses to environment; in absolute efficiency of utilizing light; in physiology; in plant dry matter production; and in efficiency of water and nitrogen utilization. These new understandings of plants have world-wide implications and applications in ecology and in agriculture. Similar work and new understandings exist and are accumulating for photosynthetic bacteria and algae; however, this chapter will emphasize these new understandings and applications with higher plants due to their world-wide agricultural importance. We will use the word efficiency in a functional manner to describe the effectiveness of plants in advantageously using photosynthesis in producing a product such as seeds or in utilizing a resource such as CO_2 or water or soil nitrogen. Though we will present and discuss such features as the biochemistry of photosynthetic CO_2 metabolism or environmental physiology separately, readers will soon realize that a specific crop or species integrates these features in a unique fashion with advantageous consequences for the plant. Thus, each crop of interest to a breeder will have some unique features influencing its efficiency which should be understood and integrated into the breeder's plant improvement work.

II. THE DIVERSITY OF PLANT PHOTOSYNTHESIS

Three to five decades ago, photosynthesis research was conducted from a somewhat restricted viewpoint of comparing photosynthetic bacteria with algae or with a few higher plants such as spinach or sunflower.[1] Our restricted view of the diversity of plant photosynthesis was broadened immeasurably by the work of the last two decades on topics such as C_4 photosynthesis, photorespiration, and environmental physiology. Indeed, we will outline six separate biochemical pathways of leaf photosynthetic CO_2 assimilation which occur in specific plants. Furthermore, we could outline in specific leaves one of two biochemical pathways of NO_3^- and NO_2^- reduction, two pathways of NH_3 assimilation, and two variations in $SO_4^=$ activation for its assimilation into cysteine. All of these variations in the assimilation of essential elements by plants require the chloroplast and are part of the diversity of photosynthesis we have learned about recently.[1]

III. BIOCHEMICAL CYCLES OF CO₂ ASSIMILATION

The six biochemical cycles of photosynthetic CO_2 assimilation which are known today occur in specific plant species. Table 1 lists the pathways along with some plants of common usage in crop, forest, or horticulture work. More exhaustive lists of species with known CO_2 fixation pathways are available.[2,3,4]

Broadly, plants have been classified as C_3, C_4, and CAM[5] but Table 1 shows that these classifications are insufficient in that specific species have quite distinct biochemical differences in CO_2 metabolism. In addition, a few species are now known which can be classified as C_3/C_4 intermediate plants. These species include *Panicum milioides*, *P. schenckii*, *P. decipiens*, *Larix leptolepis*, *Flaveria* sp., and *Moricandia* sp.[6,7,8] An aware understanding of their unique CO_2 fixation biochemistry is not available yet,[6] hence, C_3/C_4 intermediate species biochemistry is not considered further in this chapter.

Table 1
BIOCHEMICAL CYCLES OF PHOTOSYNTHETIC CO_2 ASSIMILATION IN PLANTS COMMONLY USED IN AGRICULTURE

Photosynthetic CO_2 assimilation cycle	Common crop, forest, or horticulture plants
C_3 cycle	Wheat, rice, oats, barley, all legumes, apples, pears, pines, oaks, squash, tomato, peppers, cassava, potatoes
CAM, $NADP^+$-dependent malic enzyme	*Agave* sp., *Opuntia* sp., *Kalanchoe* sp., *Sedum* sp., *Vanilla* sp., *Crassula* sp.
C_4 Cycle, $NADP^+$-dependent malic enzyme	Corn, sugarcane, *Sorghum* sp., *Paspalum* sp., *Digitaria* sp.
C_4 Cycle, PEP carbosykinase dependent	*Panicum maximum, P. texanum,* rhodesgrass
CAM, PEP carboxykinase dependent	Pineapple and other bromeliads, *Aloe* sp., *Stapelia gigantea*
C_4 Cycle, NAD^+-dependent malic enzyme	*Panicum miliaceum, Eleucine indica, Chloris distichophylla, Amaranthus* sp., *Portulaca grandiflora, Cynodon* sp.

A. The C_3 Cycle

A brief historical outline of the discovery of the six known biochemical pathways of leaf photosynthetic CO_2 assimilation is pictorially drawn in Figure 1. The first pathway of leaf CO_2 assimilation to be elucidated was the C_3 cycle (also in the literature called the reductive pentose phosphate cycle, or the photosynthetic carbon reduction cycle, or the Calvin-Benson cycle).[9,20] This well-known cycle is presented in an abbreviated form in Figure 2 which also includes the associated photorespiratory cycle (also in the literature called the glycolate pathway or the photosynthetic carbon oxidation cycle).[21] Each step in the C_3 cycle has been studied and the results are commonly available in the literature.[9,20] Indeed, the CO_2 fixation/O_2 fixation enzyme RuBP carboxylase/oxygenase is the topic of the previous chapter. And though the C_3 cycle was outlined by 1954 (Figure 1), the closely associated photorespiratory cycle (Figure 2) was not firmly elucidated until the 1970s.[21,22] Photorespiration will be discussed only briefly since the next chapter is devoted to the topic.

The C_3 cycle is widely distributed among plant species and many of the major crops that plant breeders work with exhibit the C_3 cycle (Table 1). Indeed, in each of the biochemical pathways for net leaf CO_2 assimilation in Table 1 and Figure 1, the C_3 cycle is the final step in the net reduction of carbon en route to a major storage form, e.g., starch. However, it is critical to understand that a leaf integrates all of its internal CO_2 metabolism to give a net retention of carbon. For example, in Figure 2, photorespiration is part of the total C_3 CO_2 fixation; organic acid formation and decarboxylation are integral portions of C_4 or CAM photosynthesis. Green guard cells of all plants (at least all studied to date) do not exhibit the C_3 cycle[23] even though the leaf as a whole is conducting its major photosynthesis cycle. Ultimately, the cellular CO_2 metabolism of both photosynthetic and nonphotosynthetic cells within each green leaf is integrated into the major pathway of leaf photosynthetic carbon assimilation as given in Table 1.

B. Crassulacean Acid Metabolism

Though not widely appreciated as a separate pathway of CO_2 assimilation at the time of its early study, Crassulacean acid metabolism or CAM has been known for over a century.[3,12] The biochemistry of CAM involves a net fixation of CO_2 into organic acids at night plus the operation of the C_3 cycle during the day (Figure 3).[24,25] Some of the unique biochemistry of CAM was understood by 1960[12] (Figure 1), but a much firmer understanding developed after the discovery of C_4 photosynthesis in 1965

LEAF CO₂ ASSIMILATION CYCLES

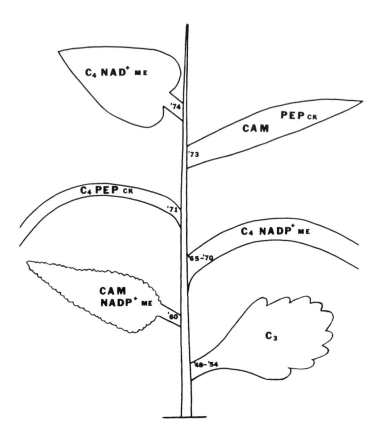

FIGURE 1.   Pictorial history of the discovery of photosynthetic CO_2 assimilation cycle in leaves. The C_3 cycle was established by 1954 following the availability of $^{14}CO_2$ after World War II.[9,10,11] Crassulacean acid metabolism (CAM) was well known in the 1950s.[12] NADP⁺-dependent malic enzyme was found in 1960[13] and PEP carboxykinase-dependent CAM plants were recognized in 1973.[14] C_4 photosynthesis was discovered in 1965—1966[15,16] followed by the recognition of NADP⁺-dependent malic enzyme C_4 plants in 1969—1970,[17] PEP carboxykinase-dependent C_4 plants in 1971,[18] and NAD⁺-dependent malic enzyme C_4 plants in 1974.[19]

to 1966.[15,16] Today, we can biochemically recognize CAM with two pathways, namely, plants using NADP⁺ malic enzyme and plants using PEP carboxykinase to decarboxylate malate and oxaloacetate, respectively (Figure 1, Table 1).

Figure 3 depicts the major unique steps in CAM which occur in each green mesophyll cell of a CAM leaf over a day. At night, starch, in most CAM plants,[12] or stored soluble sugars in pineapple,[26] are metabolized via glycolysis to form PEP. This initial CO_2 fixation reaction occurs in the cytoplasm (Figure 3). This carbon fixation results in the synthesis of malate which is stored as malic acid in the vacuole at night then released into the cytoplasm the next day for decarboxylation to produce CO_2 plus a 3-carbon fragment, either pyruvate or PEP. The light-dependent C_3 cycle of PS then operates in the CAM chloroplast using this CO_2 to produce lipids, amino acids, and other products of photosynthesis. Ultimately, the cell stores the pyruvate or PEP, formed from the respective decarboxylation of malate or oxaloacetate, either as starch or soluble sugars in stoichiometric quantities to synthesize PEP for the next night of CO_2 fixation.

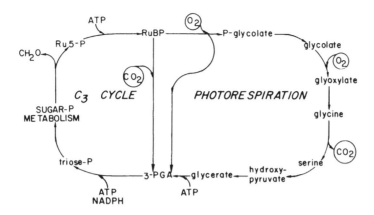

FIGURE 2. Pathway of carbon flow during the C_3 cycle of photosynthetic CO_2 fixation integrated with the pathway of carbon flow during photorespiration. The competitive carboxylation vs. oxygenation of RuBP directs carbon into these interlocking cycles. In leaves, the carbon flowing into each cycle can be modulated by varying the CO_2 or O_2 concentration.

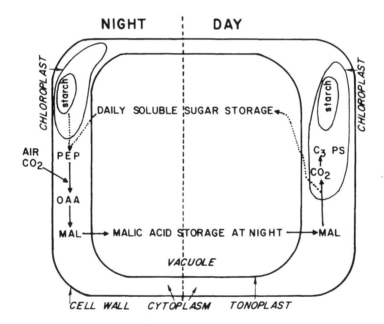

FIGURE 3. Intercellular localization of the major pathways of carbon flow in a 24-hr day during CAM. Late in a day period, if the malate is depleted, well-watered CAM plants can fix limited amounts of air CO_2 directly by the chloroplast C_3 cycle. The dotted line is a pictorial separation of the day and night periods of the metabolism of a CAM cell.

These temporal as well as spatial separations of CAM allow these plants to be extremely efficient at utilizing water. Indeed, as we will show later, they are the most efficient crops in the world in water utilization. In general, little plant breeding work has been done on CAM plants[3] even though pineapple is an important fruit crop and many horticultural plants are CAM (Table 1).

C. C_4 Cycles

Unquestionably, a major catalysis in the search for new understandings in photosynthesis rose with the discovery of C_4 photosynthesis in 1965—1966.[15,16] This discovery sparked and continues to catalyze an intensive world-wide search to uncover the diversity of photosynthesis and many other aspects of plant metabolism. Considering the 250,000 species of higher plants in the world, much remains to be learned!

C_4 photosynthesis is very effective in catalysis of research because it involves all aspects of plant biology.[4,5,27-30] A striking example is the involvement of plant anatomy; two green cell types are required for C_4 photosynthesis, as illustrated in Figure 4A. This cross section of a sorghum leaf illustrates "Kranz leaf anatomy"[27] but the essential point of understanding is that C_4 photosynthesis requires two green cell types, not necessarily Kranz, to carry out the biochemistry of photosynthesis as partially illustrated with CO_2 metabolism in Figure 4B.[29,30] Each cell type has certain biochemical reactions, e.g., PEP carboxylase in the mesophyll cell or a C_4 acid decarboxylase and RuBP carboxylase in the bundle sheath cell. Within the leaf this division of metabolism into these cells is integrated in a cooperative fashion to result in the overall fixation of CO_2 (Figure 4B).

This division of labor ultimately results in C_4 plants being efficient in utilizing CO_2, H_2O, and mineral nutrients or in avoiding O_2 inhibition of leaf metabolism. These topics will be discussed later.

As illustrated in Figure 1 and Table 1, C_4 photosynthesis is more diverse than C_3 photosynthesis since we know three separate biochemical pathways for the complete fixation of CO_2 which occur in individual plant species. The C_4 acid decarboxylase present in a given species most clearly designates these biochemical differences. The earliest work on sugarcane was with leaves containing $NADP^+$ malic enzyme which was recognized as a decarboxylase in 1969—1970 (previously, it was studied as a carboxylase). This was followed by work with *Panicum maximum* and other plants which were shown to contain PEP carboxykinase. Then *Panicum millliaceum* and other plants were shown to contain NAD^+ malic enzyme (Figure 1).

Each biochemical pathway for conducting C_4 photosynthesis has a different requirement for ATP and perhaps for reduced pyridine nucleotides in each cell type which results in differences in their light utilization (data given later). A central advantage which C_4 photosynthesis seems to confer on a leaf is that the feeding of CO_2 via biochemical intermediates directly to the bundle sheath cell chloroplast results in an efficient operation of the C_3 cycle there. The CO_2 in this cell is increased effectively and, in addition, the O_2 released from photosynthesis is less than in a C_3 leaf because part of the O_2 is evolved in the C_4 mesophyll cell where the C_3 cycle is absent (Figure 4). Thus, the competitive interlocking photorespiratory carbon pathway (Figure 2) is less active in C_4 bundle sheath cells because the O_2 is reduced and CO_2 is increased.

IV. EFFICIENCY OF PLANTS IN THE ADVANTAGEOUS UTILIZATION OF CARBON ASSIMILATION CYCLES

Having succinctly presented the six known biochemical cycles of photosynthetic CO_2 assimilation, we can now consider how efficiently plants utilize this diversity in photosynthesis to their advantage. We will first consider light utilization and then a variety of other advantageous consequences or products of these variations in photosynthetic CO_2 assimilation cycles.

A. Efficiency of Light Utilization

Only about half the irradiance of sunlight is useful in plant photosynthesis (400 to 700 nm is photosynthetical active radiation or PAR) but most agricultural plants, in fact, only convert 2 to 3% of their irradiance into the production of organic matter. Occasional reports of higher conversion percentages (7 to 8%) are rare, and sustaining

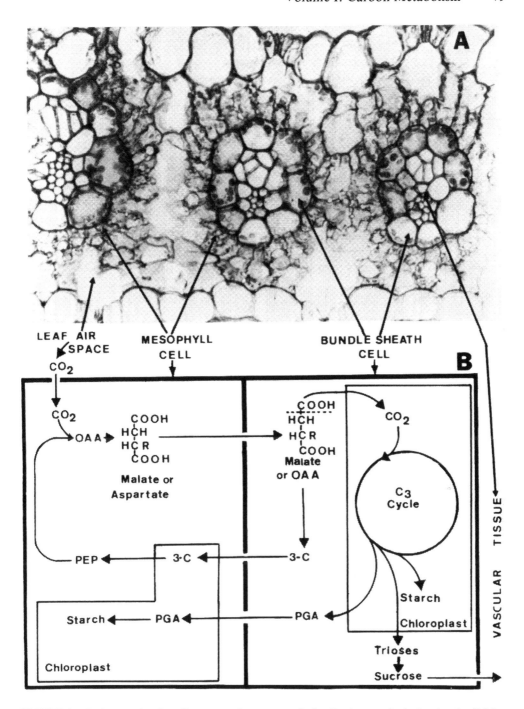

FIGURE 4.   Pathways of carbon flow to starch or sucrose during C_4 photosynthesis showing the division of metabolism into two green cell types observed in a leaf cross section. (A) Light micrograph of a sorghum leaf in cross section showing classical "Kranz leaf anatomy". Various leaf areas are labeled with arrows which correspond to the biochemical scheme. (B) Biochemical pathways of carbon flow during C_4 photosynthetic CO_2 fixation in the two cell types. This efficiently coordinated carbon flow requires two green cell types each with their designated enzymes and other required components. 3-C is the pyruvate or PEP formed in the decarboxylation of malate or oxaloacetate in specific C_4 plants (Table 1). The biochemical scheme is condensed to include all three types of C_4 photosynthesis in Table 1.

Table 2
PHOTOSYNTHETIC QUANTUM
REQUIREMENTS FOR C_3, C_4, AND CAM
PLANTS

Plant photosynthetic CO_2 assimilation cycle	Quantum requirement	Ref.
C_3, dicots	19.23[a]	32[b]
C_3, grasses	18.86[a]	32
CAM, NADP+, ME	16[c]	33
C_4, NADP+ ME, dicots	16.3	32
C_4, NADP+ ME, grasses	15.38	32
C_4, PEPCK, grasses	15.63	32
CAM, PEPCK	—	—
C_4, NAD+ ME, dicots	18.86	32
C_4, NAD+ ME, grasses	16.67	32

[a] Similar determination in 2% O_2 yield QR values near
12.8; but 2% O_2 has no influence on C_4 plant QRs.[31-33]
[b] Determined at 330 $\mu l/l$ CO_2, 21% O_2, 30°C.
[c] Measured as mole quanta/mole of malate consumed.

a high conversion percentage throughout a growing season is uncommon. As concerns general values leaves reflect or transmit about 20% of PAR and absorb 80%, though these values can vary with leaf anatomical structure and angle to the sun. The utilization of light energy by plants in agricultural systems may be constrained more by other environmental components such as water than by available light. An excess availability of light generally seems to be true in most agricultural cropping systems over the lifetime of the plant.

Even so, plants do vary in their photosynthetic ability to respond to increasing irradiance and these variations are related to the pathway of CO_2 assimilation in a specific plant. These variations occur both at light-limiting irradiations (where the absolute quantum requirement of photosynthesis is determined), and with individual leaf or plant canopy responses to increasing irradiation.

Several years ago, the available data on the quantum requirement (which is the number of light molecules or quanta required to fix one molecule of CO_2 or to evolve one molecule of O_2) for photosynthesis were evaluated in regard to the discovery of multiple pathways of CO_2 assimilation and of photorespiration.[31] Much controversy historically surrounds the quantum requirement (QR) of photosynthesis, and rightly so since the QR should accurately reflect the basic mechanisms, pathways, and efficiency of light utilization by plants. But it was clear that photorespiration and variations in pathways of CO_2 assimilation did influence the QR and these were not considered in the earlier research.[31] Specifically, if one considers the extra energy needed for photorespiration (Figure 2) or different CO_2 assimilation cycles (Figures 2, 3, and 4), the QR of plant photosynthesis should vary and be responsive to environmental changes such as O_2 or CO_2 concentrations.

In the current literature, QR values do reflect these new understandings and some of the data are collected in Table 2. In general, C_3 plants in air tend to have equal or higher QR than C_4 or CAM plants at 30°C or higher leaf temperatures (Table 2).[31-33] However, in marked contrast, the QR of C_3 plants is sensitive to temperature (20° to 40°C) and to the O_2 level (21% vs. 2%) in the atmosphere whereas the C_4 plant QR is not sensitive to either environmental change (Table 2).[31-33] The QR of C_3 plants drops to values near 12 to 14 at 2% O_2 and 20°C. This increased efficiency of light utilization certainly is a reflection of the energy requirements of photorespiration in air and its inhibition at 2% O_2.[31]

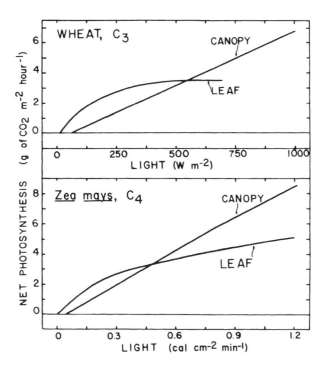

FIGURE 5. Canopy and individual leaf photosynthesis responses to increasing solar radiation to near full sunlight with a C_3 and C_4 crop. The wheat curves were adapted from original data collated by Denmead[34] in Figure 12. The wheat canopy data were collected from a crop with a leaf area index of 3.2. The maize curves were adapted from data collated by Zelitch[35] in Figures 8.1 and 9.2. The maize canopy data were collected from a crop with a leaf area index of 4.3. References to original literature and experimental datum points are given by both authors.[34,35] The canopy and leaf curves are per unit of ground and leaf surface area, respectively.

If we study the photosynthesis responses of individual leaves or of crop canopies to increasing levels of irradiance in air, we find the general responses seen in Figure 5. Similar data on CAM photosynthesis are not available. Wheat and corn are representative of C_3 and C_4 photosynthesis plants (Table 1). In both cases, canopy photosynthesis increases to full sunlight intensities at agronomic leaf area indexes. Thus, when leaf area index is such that the soil surface is well covered, light penetration to all leaves in a canopy effectively limits photosynthesis. Of course, such leaf area indexes do not occur over the lifetime of a crop.

It is in the responses of individual leaves to increasing irradiance that we find a well-recognized difference in C_3 crops vs. C_4 crops. The leaf data in Figure 5 are for an average leaf in the crop canopy; measurements made at various levels in the canopy would be above and below these curves. The C_3 leaf tends to saturate with increasing irradiance whereas the C_4 leaf does not (Figure 5). This generality was recognized soon after the discovery of C_4 photosynthesis.[4,5,27] All biologists treasure exceptions to generalities; indeed, both a C_3 and C_4 desert species is known in which leaf photosynthesis does not saturate at full sunlight.[36,37] And single leaf photosynthesis of a few C_3 crops such as cotton and sunflower tend to gradually increase to full sunlight.[35] Still, the leaf curves in Figure 5 are representative of the efficiency of light utilization by most C_3 and C_4 plants of agricultural importance and have been widely documented in the literature.[2,5,27,35]

We now know that photosynthesis in specific plants does vary in absolute efficiency or QR of light utilization (Table 2) and in the absolute amounts of carbon fixed as light is increased to full sunlight (Figure 5). This variation is related to pathway of photosynthetic CO_2 assimilation; therefore, we can begin to realize that certain advantageous benefits accrue to a plant with a given pathway of carbon assimilation.

B. Advantageous Characteristics Associated with C_3, C_4, and CAM Plant Photosynthesis

As a consequence of possessing a given pathway of photosynthesis, plants can be more efficient in utilizing their environmental resources in addition to light. Over a decade ago, I collected about 20 characteristics which distinguished three photosynthetic groups of higher plants[5] and some of these plus more recently discovered characteristics have advantages for plants apparently as a direct consequence of their photosynthetic carbon assimilation cycle.

Table 3 summarizes characteristics related to the efficiency of growth or resource utilization by plants either with C_3 or C_4 or CAM photosynthesis. Less data are available on CAM plants, so mainly we will compare C_3 with C_4 plants. However, insufficient data are available to compare types of C_4 or CAM plants (as in Tables 1 or 2) with C_3 plants.

In assessing the maximum short-time growth rates of plants, the literature must be read and interpreted in view of the use of good experimental procedures in the field, e.g., the avoidance of plot edge effects. Monteith has discussed many of these problems and analyzed the literature data on growth rates for C_3 and C_4 crops.[38] Table 3 presents a summary of the maximum growth rate showing the 34 to 39 and 51 to 54 gm of dry weight produced per m^2 of ground surface area per day for four C_3 and three C_4 crops.

Maximum short-time growth rates should be reflected in the yield of a crop; however, it is difficult to maintain optimum plant growth conditions throughout the lifetime of a crop. In addition, crops have various lifetimes. So, in a perhaps more useful analysis, Monteith calculated the mean annual growth rate for a number of common C_3 and C_4 crops. These data are presented in more detail in Figure 6 which is drawn to compensate for various growing seasons of individual crops. Clearly, the dry weight of C_4 crops consistently is higher than C_3 crops independent of the crop growing season.

The slopes of the two lines on Figure 6 are the mean annual or crop season growth rates (Table 3). These C_4 crops have about a 70% higher mean annual growth rate than the C_3 crops which is reflected in the approximately 40% higher crop dry weight at harvest values (Figure 6). Therefore, either from maximum short-time and mean seasonal growth rates or from crop yields (Table 3, Figure 6), C_4 crops are different from C_3 crops in an important agricultural characteristic, namely, yield, and these differences are a reflection of the higher leaf photosynthesis rates of C_4 crops (Table 3).

CAM plants often grow in dry agricultural systems and their reported growth and photosynthesis rates are low (Table 3) as is dry matter production. But in Hawaii, pineapple yields can be quite comparable to C_3 plants[39] (Table 3). The most striking advantage a CO_2 fixation pathway gives to a plant probably is the water use efficiency of CAM plants (Table 3). This surely is a reflection of their major carbon assimilation occurring at night (Figure 3)[3,12,24,25,39] followed by daytime stomatal closure which greatly reduces leaf transpiration through the day when the irradiance intensity is greatest.

Another clearcut advantage accruing to a plant based on its primary cycle of CO_2 assimilation is the efficient use by C_4 plants of air levels of CO_2 along with no (or little) O_2 inhibition of leaf photosynthesis (Table 3). The efficient use of CO_2 reflects the biochemical capability of the leaf PEP carboxylase (Figure 4) in fixing CO_2 (indeed,

Table 3

ADVANTAGEOUS CHARACTERISTICS DISTINGUISHING PLANTS WITH
SEPARATE CYCLES OF PHOTOSYNTHETIC CO_2 ASSIMILATION

Advantageous characteristic[a]	Primary cycle of CO_2 assimilation			
	C_3	C_4	CAM	Ref.
Maximum short-time growth rate:				
gm of dry wt/dm^2 of leaf area/day	0.5 to 2	4 to 5	0.015 to 0.018	5[b]
gm/m^2 of ground/day	34 to 39	51 to 54	—	38
Mean annual growth rate:				
gm dry wt/m^2 of ground/day	13 ± 1.6	22 ± 3.6	—	38
Dry matter production:				
Harvested tons/hectare/year	∿45	∿75	Generally 2 to 8; maximum 44	3, 38, 39
Transpiration ratio:				
gm H_2O used/gm of dry wt produced	450 to 950	250 to 350	18 to 155	3, 5
Maximum rate of net photosynthesis:				
mg of CO_2/dm^2 of leaf surface/hr	15 to 40	40 to 80	Generally 1 to 4; maximum 11 to 15	3, 5
Optimum day temperature, range for net CO_2 assimilation:	15 to 30°C	30 to 45°C	<25°C	3, 5
CO_2 saturation concentration for leaf photosynthesis:				
μl of CO_2/l of air	>600 to 800	near 400	>600 to 800	27, 40
O_2 Inhibition of leaf photosynthesis:				
2% O_2 vs. 21%	35 to 40%	<5%	20 to 45%[d]	5, 40
Nitrogen use in:[e]				
leaf photosynthesis rate, mg CO_2/dm^2/hr/% leaf N;	4.7 to 5.6	9 to 10	—	41, 42
Kg dry matter produced/ Kg of fertilizer N	34	74	—	30

[a] Comparisons are made in air (21% O_2, 340 μl of CO_2/l of air) and other under as near healthy physiological conditions as possible. A dash indicates data could not be located.

[b] Only selected references are cited for interested readers to have a literature beginning on each characteristic.

[c] Net CO_2 fixation in CAM plants during the day is strongly influenced by the previous night temperature. Higher temperatures generally reduce net day CO_2 uptake.[3]

[d] 4% O_2 vs. 20%, late in the day after acid depletion.

[e] Most of these results were obtained with grasses.

PEP carboxylase can reduce CO_2 to zero in a closed container) and the sequestering of RuBP carboxylase in the interior of the leaf. In the bundle sheath cells, the CO_2 level is effectively raised and O_2 is reduced so that neither CO_2 is rate limiting for their C_3 cycle nor does the competition by O_2 (Figure 2) occur in amounts sufficient to effectively reduce CO_2 fixation.

No doubt these characteristics are related in turn to the ability of C_4 plants to utilize higher temperatures for growth and photosynthesis and to efficiently use H_2O in dry matter production (Table 3). Transposed into our agricultural cropping systems, these characteristics mean that, in temperate agriculture, C_4 plants are summer annual crops such as maize and sorghum or perennials such as bermudagrass or weeds such as crabgrass, and the winter crops are C_3 plants. In tropical agriculture, C_4 plants are major crops and/or dominate weeds in the C_3 crops.

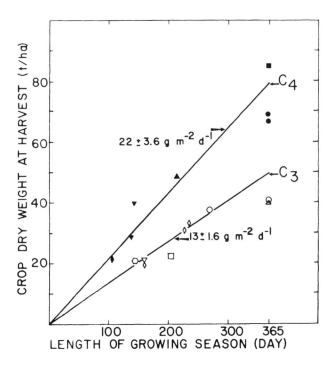

FIGURE 6.   Crop dry weight at harvest in tons per hectare vs.
the length of the crop growing season in days for C_3 and C_4 crops.
Figure modified from data collated by Monteith.[38] The solid symbols are for C_4 crops and the open symbols are for C_3 crops. The
C_4 crops are bulrush millet, maize, sorghum, sugarcane, and na-
pier grass; and the C_3 crops are kale, potatoes, sugar beet, rice,
cassava, and oil palm. The line drawn through each set of crops is
the mean crop yield vs. growing season and, from its slope, Monteith calculated a mean seasonal growth rate in grams of crop dry
weight produced per square meter of land surface per day.

C_4 photosynthesis requires two green cell types in which some proteins are unique to
each cell type.[30] In other words, each photosynthetic cell type does not synthesize every
enzyme needed for net CO_2 assimilation. This division of labor results in an efficient
use of leaf N, and of soil N as well.[41] Brown recognized that C_4 plants made more
efficient use of leaf N than C_3 plants in both dry matter production and in catalysis of
photosynthesis.[41,42] He demonstrated that leaf photosynthesis is a linear function of
leaf N content and has calculated that leaf photosynthesis per unit of leaf N (% leaf
N) is nearly twofold higher in C_4 plants than in C_3 plants (Table 3). Or, if one measures
dry matter production per unit of fertilizer N added, about twofold more dry matter is
produced by C_4 plants (Table 3). In modern agriculture, the efficient utilization of N-
fertilizer is an advantageous characteristic.

With such a list of advantageous characteristics accruing to a plant based on its
primary cycle of CO_2 assimilation (Table 3), one would conclude that this knowledge
has useful application in agriculture. Let us examine this thesis.

V. IMPLICATIONS AND APPLICATIONS OF PHOTOSYNTHESIS IN AGRICULTURE

In the absence of photosynthesis, crop yield is zero! This axiom, however, has not
often resulted in our accumulating knowledge about the mechanisms or the diversity

of photosynthesis being directly applied in agriculture in increasing crop fiber, forage, or tree yields. Another axiom is the dependency of animal life upon photosynthesis. Why, then, hasn't knowledge about photosynthesis been more widely used in plant breeding work or in plant production?

No single answer is adequate but one limitation rests in the problems associated with measuring photosynthesis, particularly on a scale comparable to most plant breeding work which often involves many crosses where the parents and progeny need to be measured. Attempts were made with several crops over the last three decades to measure leaf photosynthesis and to correlate this measurement with crop yield. To my knowledge, most of these attempts were futile. The problem, in most cases, was that photosynthesis was measured at a single (or only a few) time(s) with a single leaf or a leaf section at one or only a few stages of plant development. These single or limited number of measurements then were correlated with yield. Usually, yield was seed production. Seed yield or growth of a plant is a complex integration over the lifetime of a plant of its genetic capability plus the environmental restrictions experienced.[43] Thus, in retrospect, it seems unlikely that single measurements of the photosynthetic activity of a plant would correlate with yield or growth which is an integration of much of its lifetime. The question about the limited application of photosynthesis research in agriculture could be asked differently, assuming good cultural practices are employed: does photosynthesis limit yield?[43]

In several recent reviews, Evans[44] and Gifford[46] found little evidence for a positive correlation between photosynthetic activity and yield. Nor have they found evidence that the selection of high yielding crops, e.g., wheat, sorghum, corn, sugarcane, pearl millet, cotton, or cowpeas, by breeders has been a result of increased rates of leaf photosynthesis. Indeed, leaf photosynthesis activity is higher for the wild relatives of wheat, sorghum, pearl millet, and cotton than for their modern cultivars.[46] However, other work shows that modern cultivars of peanuts have higher leaf photosynthesis rates than wild *Arachis* species.[47]

In a different approach by measuring canopy photosynthesis, a recent report shows a strong relationship between yield of soybeans and seasonal net photosynthesis.[48] In these studies covering several years of field research, canopy photosynthesis was measured regularly over a day and throughout the growing season. Total sunlight irradiance also was measured. By obtaining seasonal photosynthesis and plotting it vs. seed yield, a linear relation was obtained. Figure 7 summarizes these studies for soybeans using several cultivars, planting dates, and methods for varying crop density such as row and drill spacing. Clearly, soybean yield is a linear function of seasonal photosynthesis under a variety of field conditions.

Two direct conclusions regarding yield increases can be drawn from Figure 7: first, yield can be increased if net seasonal photosynthesis is increased; or second, yield can be increased if the steepness of the line is increased. A third conclusion from this linear response is that soybean harvest index is constant. In other words, an increase in the slope of the line is equal to an increase in harvest index. Figure 7 is intuitively satisfying and quite informative; but obtaining seasonal canopy photosynthesis is very laborious for most plant breeding programs and the determinants or controls of harvest index are quite uncertain. Breeding for high harvest index unquestionably has been accomplished in the past and will continue to be an invaluable crop breeding improvement tool.[49] Finally, canopy photosynthesis data are not available for other crops; therefore, extrapolation must be done with caution.

Boerma and associates,[50,51] in a combined breeding and physiology research program, have studied canopy photosynthesis and yield with determinate soybean varieties. They find positive correlations between canopy photosynthesis and seed yield sufficiently high to be used in selection for yield. But much more labor was required to obtain canopy photosynthesis than seed yield.

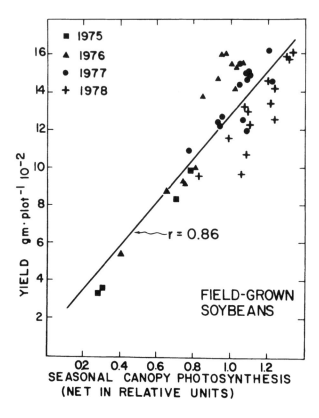

FIGURE 7.   Soybean grain yield vs. net seasonal canopy photosynthesis. Net photosynthesis was obtained from regular daily
measurements of canopy photosynthesis with field-grown soybeans. The data include two cultivars, four growing seasons, and
treatments such as row spacing, planting density, and planting
date. All points are means of six field plots. The solid line is a
computer fitted line showing a regression coefficient of r = 0.86.
Data of Dr. L. A. Christy; figure modified from Christy and Porter.[48]

Other types of work also have answered the question, does photosynthesis limit
yield. For example, in a variety of C_3 crops, CO_2 enrichment increases photosynthesis
and yield as does the addition of supplemental lighting under a dense canopy or the
practice of skip-row planting.[46,51]

VI. CONCLUSIONS

Higher plants, including many agriculturally useful plants, have evolved six known
biochemical cycles for photosynthetic CO_2 assimilation. One cycle is the dominant
pathway for CO_2 assimilation in a specific crop. As a result of possessing a specific
cycle of CO_2 assimilation, several advantageous consequences accrue to a plant. Thus,
C_3, C_4, and CAM plants all have advantages in specific environments which make them
productive and efficient in the utilization of the resources of the environment. Net
seasonal canopy photosynthesis is positively correlated with seed yield in soybeans.
Canopy photosynthesis is light limited at high leaf area indexes. Selection for high total
canopy photosynthesis combined with selection for high harvest index is recommended
for increasing yields in plant breeding improvement work.

REFERENCES

1. Black, C. C., Biosolar resources: fundamental biological processes, in *CRC Handbook of Biosolar Resources,* Vol. 1 (Part 1), Mitsui, A. and Black, C. C., Eds., CRC Press, Boca Raton, Fla., 1982, 3.
2. Smith, B. N., General characteristics of terrestrial plants (agronomic and forests) — C_3, C_4, and Crassulacean Acid metabolism plants, in *CRC Handbook of Biosolar Resources,* Vol. 1 (Part 2), Mitsui, A. and Black, C. C., Eds., CRC Press, Boca Raton, Fla., 1982, 99.
3. Kluge, M. and Ting, I. P., *Crassulacean Acid Metabolism (Analysis of an Ecological Adaptation),* Springer-Verlag, Berlin, 1978.
4. Burris, R. H. and Black, C. C., *Carbon Dioxide Metabolism and Plant Productivity,* University Park Press, Baltimore, 1976.
5. Black, C. C., Photosynthetic carbon fixation in relation to net CO_2 uptake, *Annu. Rev. Plant Physiol.,* 24, 253, 1973.
6. Brown, R. H., Bouton, J. H., Rigsby, L., and Rigler, M., Photosynthesis of grass species differing in carbon dioxide fixation pathways. VIII. Ultrastructural characteristics of *Panicum* species of the *Laxa* group, *Plant Physiol.,* 71, 425, 1983.
7. Fry, D. J. and Phillips, I. D. J., Photosynthesis of conifers in relation to annual growth cycles and dry matter production, *Physiol. Plant.,* 37, 185, 1976.
8. Apel, P. and Maass, I., Photosynthesis in species of *Flaveria.* CO_2 compensation concentration, O_2 influence on photosynthetic gas exchange and $d^{13}C$ values in species of *Flaveria* (Asteraceae), *Biochem. Physiol. Pflanz.,* 197, 396, 1981.
9. Bassham, J. A. and Calvin, M., *The Path of Carbon in Photosynthesis,* Prentice-Hall, Englewood Cliffs, N.J., 1957.
10. Norris, L., Norris, R. E., and Calvin, M., A survey of the rates and products of short-term photosynthesis in plants of nine phyla, *J. Exp. Bot.,* 6, 64, 1955.
11. Bassham, J. A., Photosynthetic carbon dioxide assimilation via the reductive pentose phosphate cycle (C_3 cycle), in *CRC Handbook of Biosolar Resources,* Vol. 1 (Part 1), Mitsui, A. and Black, C. C., Eds., CRC Press, Boca Raton, Fla., 1982, 181.
12. Ranson, S. L. and Thomas, M., Crassulacean acid metabolism, *Annu. Rev. Plant Physiol.,* 11, 81, 1960.
13. Walker, D. A., Physiological studies on acid metabolism. VII. Malic enzyme from *Kalanchoe crenata:* effects of carbon dioxide concentration, *Biochem. J.,* 74, 216, 1960.
14. Dittrich, P., Campbell, W. G., and Black, C. C., Phosphoenolpyruvate carboxykinase in plants exhibiting Crassulacean acid metabolism, *Plant Physiol.,* 52, 357, 1973.
15. Kortschak, H. P., Hartt, C. E., and Burr, G. E., Carbon dioxide fixation in sugarcane leaves, *Plant Physiol.,* 40, 209, 1965.
16. Hatch, M. D. and Slack, C. R., Photosynthesis by sugarcane leaves. A new carboxylation reaction and the pathway of sugar formation, *Biochem. J.,* 101, 103, 1966.
17. Andrews, T. J., Johnson, H. S., Slack, C. R., and Hatch, M. D., Malic enzyme and aminotransferases in relation to 3-phosphoglycerate formation in plants with the C_4-dicarboxylic acid pathway of photosynthesis, *Phytochemistry,* 10, 9, 1971.
18. Edwards, G. E., Kanai, R., and Black, C. C., Phosphoenolpyruvate carboxykinase in leaves of certain plants which fix CO_2 by the C_4-dicarboxylic acid cycle of photosynthesis, *Biochem. Biophys. Res. Commun.,* 45, 278, 1971.
19. Hatch, M. D. and Kagawa, T., NAD malic enzyme in leaves with C_4-pathway photosynthesis and its role in C_4 acid decarboxylation, *Arch. Biochem. Biophys.,* 160, 346, 1974.
20. Calvin, M., The pathway of carbon in photosynthesis, *Science,* 135, 879, 1962.
21. Tolbert, N. E., Photorespiration, in *The Biochemistry of Plants,* Vol. 2, Davies, D. O., Ed., Academic Press, New York, 1980.
22. Lorimer, G. H., The carboxylation and oxygenation of ribulose 1,5-bisphosphate: the primary events in photosynthesis and photorespiration, *Annu. Rev. Plant Physiol.,* 32, 349, 1981.
23. Outlaw, W. H., Jr., Carbon metabolism in guard cells, *Recent Adv. Phytochemistry,* 16, 185, 1982.
24. Black, C. C., Crassulacean acid metabolism, in *CRC Handbook of Biosolar Resources,* Vol. 1 (Part 1), Mitsui, A. and Black, C. C., Eds., CRC Press, Boca Raton, Fla., 1982, 191.
25. Ting, I. P. and Gibbs, M., Eds., *Crassulacean Acid Metabolism,* American Society of Plant Physiologists, Rockville, Md., 1982, 1.
26. Black, C. C., Carnal, N. W., and Kenyon, W. H., Compartmentation and the regulation of CAM, in *Crassulacean Acid Metabolism,* Ting, I. P. and Gibbs, M., Eds., American Society of Plant Physiologists, Rockville, Md., 1982, 51.
27. Hatch, M. D., Osmond, C. B., and Slatyer, R. O., Eds., *Photosynthesis and Photorespiration,* Wiley-Interscience, New York, 1971, 1.

28. Black, C. C., Chen, T. M., and Brown, R. H., Biochemical basis for plant competition, *Weed Sci.*, 17, 338, 1969.
29. Ray, T. B. and Black, C. C., The C_4 pathway and its regulation, in *Photosynthesis II. Photosynthetic Carbon Metabolism and Related Processes*, Gibbs, M. and Latzko, E., Eds., Encyclopedia of Plant Physiology, New Series, Vol. 6, Springer-Verlag, New York, 1979, 77.
30. Campbell, W. H. and Black, C. C., Cellular aspects of C_4 leaf metabolism, *Recent Adv. Phytochemistry*, 16, 223, 1982.
31. Campbell, W. H. and Black, C. C., The relationship of CO_2 assimilation pathways and photorespiration to the physiological quantum requirement of green plant photosynthesis, *BioSystems*, 10, 253, 1978.
32. Ehleringer, J. and Percy, R. W., Variation in quantum yield for CO_2 uptake among C_3 and C_4 plants, *Plant Physiol.*, in press, 1983.
33. Spalding, M. H., Edwards, G. E., and Ku, M. S. B., Quantum requirement for photosynthesis in *Sedum praealtum* during two phases of Crassulacean acid metabolism, *Plant Physiol.*, 66, 463, 1980.
34. Denmead, O. T., Temperate cereals, in *Vegetation and The Atmosphere*, Vol. 2, Monteith, J. L., Ed., 1976, 1.
35. Zelitch, I., *Photosynthesis, Photorespiration, and Plant Productivity*, Academic Press, New York, 1971, 1.
36. Björkman, O., Pearcy, R. W., Harrison, T., and Mooney, H., Photosynthesis adaptation to high temperatures: a field study in Death Valley, California, *Science*, 175, 786, 1972.
37. Mooney, H. A., Ehleringer, J., and Berry, J. A., High photosynthetic capacity of a winter annual in Death Valley, *Science*, 194, 322, 1976.
38. Monteith, J. L., Reassessment of maximum growth rates for C_3 and C_4 crops, *Exp. Agric.*, 14, 1, 1978.
39. Marzola, D. L. and Bartholomew, D. P., Photosynthetic pathway and biomass energy production, *Science*, 205, 555, 1979.
40. Osmond, C. B. and Björkman, O., Pathways of CO_2 fixation in the CAM plant *Kalanchoe daigremontiana*. II. Effects of O_2 and CO_2 concentration on light and dark CO_2 fixation, *Aust. J. Plant Physiol.*, 2, 155, 1975.
41. Brown, R. H., A difference in N use efficiency in C_3 and C_4 plants and its implications in adaptation and evolution, *Crop Sci.*, 18, 93, 1978.
42. Bouton, J. H. and Brown, R. H., Photosynthesis of grass species differing in carbon dioxide fixation pathways. V. Responses of *Panicum maximum, Panicum milioides* and tall fescue (*Festuca arundinacea*) to nitrogen nutrition, *Plant Physiol.*, 66, 97, 1980.
43. Black, C. C., Sunbeams, pathways and plants, *Antioch Rev.*, 38, 436, 1980.
44. Gifford, R. M., A comparison of potential photosynthesis, productivity and yield of plant species with differing photosynthetic metabolism, *Aust. J. Plant Physiol.*, 1, 107, 1974.
45. Evans, L. T., Physiological adaptation to performance as crop plants, *Philos. Trans. R. Soc.*, London Ser. B, 275, 71, 1976.
46. Gifford, R. M. and Evans, L. T., Photosynthesis, carbon partitioning, and yield, *Annu. Rev. Plant Physiol.*, 32, 485, 1981.
47. Bhagsari, A. S. and Brown, R. H., Photosynthesis in peanut (*Arachis*) genotypes, *Peanut Sci.*, 3, 1, 1976.
48. Christy, L. A. and Porter, C. A., Canopy photosynthesis and yield in soybean, in *Photosynthesis: Development, Carbon Metabolism, and Plant Productivity*, Vol. 2, Govindjee, Ed., Academic Press, New York, 1982, 499.
49. Donald, C. M. and Hamblin, J., The biological yield and harvest index of cereals as agronomic and plant breeding criteria, *Adv. Agron.*, 28, 361, 1976.
50. Harrison, S. A., Boerma, H. R., and Ashley, D. A., Heritability of canopy-apparent photosynthesis and its relationship to seed yield in soybeans, *Crop Sci.*, 21, 222, 1981.
51. Wells, R., Schulze, L. L., Ashley, D. A., Boerma, H. R., and Brown, R. H., Cultivar differences in canopy apparent photosynthesis and their relationship to seed yield in soybeans, *Crop Sci.*, 22, 886, 1982.

Chapter 7

REGULATION OF PHOTORESPIRATION

C. R. Somerville and S. C. Somerville

TABLE OF CONTENTS

I. INTRODUCTION

One of the most powerful tools in science is the "working model". Such devices provide structure to a collection of otherwise isolated observations, and frequently stimulate discovery by permitting extrapolation and interpolation. Equally important, perhaps, a good model may permit a process or procedure to be exploited long before the details have been completely assembled. In this respect, knowledge of photorespiration has arrived a level of maturity where, although certain details remain obscure, a predictive model exists. This model evolved from a series of discoveries following the recognition that apparent or net photosynthesis is a balance between CO_2-fixation and a concurrent light-dependent release of CO_2 termed photorespiration.[1] First, the connection between glycolate metabolism and photorespiration was established.[2] This was followed by elucidation of the pathway by which glycolate is metabolized, and the distribution of relevant enzymes in several organelles.[3] Finally, the mechanism of glycolate synthesis was discovered[4,5] and the biochemical and gas exchange data reconciled in a coherent and predictive model.[6]

In its most abbreviated form, the model proposed by Laing et al.[6] states that photorespiratory phenomena can be explained on the basis of the properties of a single enzyme — ribulose-1,5-bisphosphate carboxylase/oxygenase (henceforth abbreviated RubisCO). The degree to which this model accounts for photorespiratory phenomena, and is supported by experimental evidence, has been the subject of several cogent reviews.[7-12]

In this article, we have avoided, as much as possible, a repetitious description of the argument in favor of the RubisCO-based model and attempted instead to explore the implications of this model in the context of attempts to reduce photorespiration in C_3 species. We have tried to describe why we consider the objective worth pursuing, what some of the technical or conceptual problems and pitfalls are, where the emphasis should be placed in future research, and what the possibilities may be for eventual success. We have avoided the temptation of referring only to the subset of published literature which supports our view and have included critical reference to arcane or apparently erroneous reports which may otherwise be somewhat confusing to the neophyte. Because of space limitations, we have emphasized the recent literature, and have avoided a discussion of photorespiration in C_4 species — a topic which has been recently reviewed.[13]

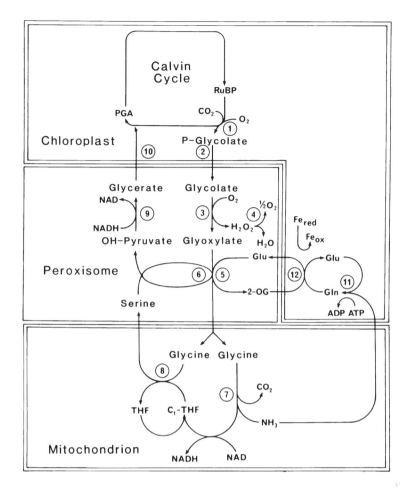

FIGURE 1. The photorespiratory pathway. Because of the two complementary routes of glycine metabolism in the mitochondrion, two molecules of P-glycolate must enter the pathway for each molecule of serine, CO_2, and NH_3 produced. The circled numbers correspond to the following enzymes: (1) RubisCO; (2) P-glycolate phosphatase; (3) glycolate oxidase; (4) catalase; (5) glutamate:glyoxylate aminotransferase; (6) serine:glyoxylate aminotransferase; (7) glycine decarboxylase; (8) serine transhydroxymethylase; (9) hydroxypyruvate reductase; (10) glycerate kinase; (11) glutamine synthetase; (12) ferredoxin-dependent glutamate synthase. The abbreviations include C_1THF, N^5, N^{10}-methylene tetrahydrofolic acid; Fe_{ox}, Fe_{red}, oxidized and reduced ferredoxin, respectively; gln, glutamine; glu, glutamate; 20 G, 2-oxoglutarate; PGA, 3-phosphoglycerate; RuBP, ribulose-1,5-bisphosphate; THF, tetrahydrofolic acid.

II. A PREDICTIVE MODEL OF PHOTORESPIRATION

The biochemical basis for many of the phenomena associated with photosynthetic and photorespiratory carbon metabolism in C_3 plants can be accounted for by the scheme presented in Figure 1. According to this scheme,[3,7] photosynthetic and photorespiratory carbon metabolism comprise two interlocking cycles which are initiated by the carboxylation or oxygenation of ribulose-1,5-bisphosphate (RuBP). Both reactions are catalyzed by the bifunctional chloroplast enzyme RubisCO.[5,14,15] Carboxylation of RuBP produces two molecules of 3-phosphoglyceric acid (PGA) which are subsequently utilized in the reactions of the Calvin cycle to regenerate RuBP. Excess fixed carbon is stored in the chloroplast as starch or exported to the cytoplasm as triose phosphate. Oxygenation of RuBP produces one molecule each of PGA and 2-phosphoglycolate (P-glycolate).

Carbon dioxide and O_2, the substrates of the carboxylase and oxygenase reactions, respectively, behave as linearly competitive inhibitors of the oxygenase and carboxylase activities, respectively. Laing et al.[6] showed that since $Km (CO_2) \simeq Ki (CO_2)$ and $Km (O_2) \simeq Ki (O_2)$ for RubisCO, the rates of carboxylase (v_c) and oxygenase (v_o) reactions can be expressed by the equations:

$$v_c = \frac{Vmax(CO_2)}{1 + \dfrac{Km(CO_2)}{pCO_2}\left[1 + \dfrac{pO_2}{Km(O_2)}\right]} \tag{1}$$

$$v_o = \frac{Vmax(O_2)}{1 + \dfrac{Km(O_2)}{pO_2}\left[1 + \dfrac{pCO_2}{Km(CO_2)}\right]} \tag{2}$$

where $Km (CO_2)$ and $Km (O_2)$ are the apparent Michaelis-Menten constants for CO_2 and O_2, and where pCO_2 and pO_2 are substrate concentrations of CO_2 and O_2, respectively. Photosynthesis is then given by the equation:

$$p = v_c - kv_o \tag{3}$$

where k is the proportion of P-glycolate carbon released as photorespiratory CO_2 by the subsequent metabolism of P-glycolate.

By substituting kinetic constants derived from in vitro studies and intracellular gas concentrations into the equations.[6,7,17,18] it can be seen that O_2 exhibits an inhibitory effect on net photosynthesis due to two distinguishable effects. First, O_2 competitively reduces carboxylation. Second, O_2 increases the amount of oxygenase activity and, therefore, the amount of CO_2 released during the metabolism of P-glycolate. This reduces apparent photosynthesis by partially negating the effect of concurrent carboxylation. By a similar analysis of the effects of varying CO_2 concentrations it can be seen that photosynthesis is stimulated by CO_2 concentrations greater than the normal atmospheric levels for two reasons. First, a high pCO_2 competitively reduces v_o. Second, since the CO_2 concentration in the liquid phase of the cell (5 to 6 μM) is about half the average value for $Km (CO_2)$, increasing the pCO_2 stimulates carboxylation by decreasing the degree of substrate limitation. Because of this latter effect, the rates of photosynthesis obtained in low O_2 are not as high as the rate obtained at saturating CO_2.

Farquhar[17,18] has described elaborations of this oversimplified model which take into account the dependence on RuBP concentration, the activation of both catalytic activities by CO_2, and the temperature dependencies of both reactions. These more detailed formulations are useful for examining the molecular basis of photosynthetic response to temperature, light intensity, and atmospheric composition.

Before exploring the implications of this model, it may be worth reiterating that O_2 has two distinguishable but intimately related effects which reduce net CO_2-fixation: stimulation of P-glycolate production with subsequent release of CO_2 as photorespiration, and reduction of apparent photosynthesis by competitive inhibition of carboxylation. Since, in principle, one may contemplate a reduction in the amount of photorespiration without any reduction in O_2 inhibition, it is useful to retain this distinction which is often blurred in common usage.

III. AGRONOMIC SIGNIFICANCE

A. Effects of CO_2 Enrichment on Growth and Yield

One of the primary reasons that the control of photorespiration has attracted attention as a possible means of improving plant productivity is that it is possible to artificially reduce photorespiration and O_2-inhibition of photosynthesis in experimental situations which also permit an assessment of the effects on growth and yield. This is possible because the magnitude of both effects is proportional to pO_2 concentration and inversely proportional to pCO_2 concentration. Thus, the effect of genetically reducing RuBP oxygenase activity may be simulated by altering the O_2 and/or CO_2 concentration of the ambient gas regime in a confined area and measuring the effects of the resulting alteration in photorespiration and photosynthesis on growth and yield.

In an early experiment of this type, Cooper and Brun[19] exposed two soybean varieties to 350 or 1350 μl CO_2 l^{-1} from seedling to maturity stages in Mylar chambers in a greenhouse. As a result of the elevated pCO_2, seed yields were increased by 51% in one variety, and by 40% in another. In shorter CO_2-enrichment studies on field-grown enclosed soybeans at different stages of development, yields were increased up to 43% depending on the state of development.[20] Krenzer and Moss[21] found that 600 μl CO_2 l^{-1} applied to field-grown wheat during floral initiation or grain filling stages increased yields by 15% in one variety and 38% in another. Other studies with wheat,[22-24] barley,[25] and soybeans[26] have uniformly shown enhanced yield due to CO_2-enrichment. CO_2-enrichment has also been reported to dramatically enhance N_2-fixation by nodulated soybeans[27] although the mechanism is still under investigation.[28]

These experiments demonstrate quite clearly that long-term enhancement of net photosynthesis is strongly correlated with increased yield. This conclusion should be distinguished from the widely accepted notion that measurement of photosynthetic rate is not correlated with yield.[29,30] Although in a strict sense these experiments do not directly simulate the suppression of photorespiration and O_2-inhibition, the net effect of a stimulation of photosynthesis by this means is expected to be equivalent to that obtained by CO_2-enrichment.

One caveat which must be appended to the foregoing argument is that at least some species appear unable to respond positively to CO_2-enrichment of indefinite duration.[29] Studies with soybean,[31] tomato,[32] and wheat[33] have indicated that after prolonged CO_2-enrichment, plants may undergo morphological and/or biochemical adjustments which partially or completely negate or obscure the effects of CO_2-enrichment. These observations may provide a partial explanation for the interaction between stage of growth at which CO_2-enrichment occurs and the extent of yield enhancement. Such observations also emphasize the point that efforts to increase productivity by enhancing photosynthesis may ultimately necessitate selection for plants that can effectively utilize enhanced rates of photosynthesis.[30]

B. Effects of O_2-Depletion on Growth

The complementary approach to CO_2-enrichment studies is to lower the O_2 concentration. At normal levels of CO_2, a decrease in the O_2 concentration from 21% to 2% causes an approximate 50% increase in net photosynthesis rate because of decreased photorespiration and O_2-inhibition.[7] The growth of young *Phaseolus vulgaris* seedlings was doubled in 2.5% O_2 vs. 21% O_2, whereas the growth of *Z. mays* (a C_4 species) was relatively unaffected.[34,35] Similar effects have been observed with several other species under similar circumstances.[36,37] Because of the substantial cost of maintaining plants under low O_2 concentrations, there have been no large-scale trials of the effects of O_2 on economic yield. Such studies might, in any event, prove somewhat misleading because of the observation that reduced O_2-concentrations may inhibit seed maturation.[38]

Although the precise mechanism of this O_2 effect has not been elucidated, it is considered to be unrelated to photosynthesis or photorespiration. Thus, although evaluation of O_2 effects on seed yield are complicated by other factors, it seems clear that enhancement of photosynthesis by *either* CO_2-enrichment or O_2-depletion results in stimulation of growth. The implication is that enhancement of net photosynthesis by genetic reduction of photorespiration and O_2-inhibition of photosynthesis would result in growth enhancement of similar magnitude.

IV. PHOTORESPIRATORY GAS EXCHANGE

Any attempt to identify potentially useful genetic variation in the amount of photorespiration or O_2-inhibition of photosynthesis is ultimately dependent upon an accurate measure of these effects. Although other indicators have been employed for some purposes, the final criteria are gas exchange characteristics, usually determined by infrared gas analysis or mass spectrometry.

Photorespiration was originally observed as a transient burst of CO_2 evolution which occurred immediately after leaves were darkened.[1] This effect, now called the post-illumination burst (PIB), was correctly interpreted as evidence that illuminated leaves were simultaneously evolving (respiring) and fixing CO_2[1], and that when the light was interrupted, fixation declined more rapidly than respiration thereby transiently revealing the "photorespiratory" CO_2-efflux. Since then, a number of methods of directly or indirectly observing photorespiration have been devised. However, most or all of these methods are of uncertain quantitative accuracy. Because photosynthesis, which is characterized by CO_2 uptake and O_2 evolution, and photorespiration (which is of opposite effect) occur simultaneously in standard atmospheric conditions, neither process can be accurately measured by net gas exchange determinations. Photosynthesis is diminished by concurrent photorespiration, and photorespiration is obscured by photosynthesis. For this reason, methods of assessing photorespiration commonly employ artificial conditions which suppress photosynthesis, or indirect measures which assess the influence of photorespiratory activity on photosynthesis. Only two methods, involving simultaneous measurement of $^{14}CO_2/^{12}CO_2$ or $^{18}O_2/^{16}O_2$, allow relatively direct measurement of photorespiration under conditions of steady-state photosynthesis.

Before describing the various gas exchange methods that have been devised to measure photorespiration, it may be useful to clarify some of the commonly used terminology.[8,13] *True photosynthesis* represents the gross amount (or rate) of CO_2-fixation. *Apparent photosynthesis* is the measured rate of photosynthesis. *Apparent photorespiration* is the difference between these two values. Those unfamiliar with gas exchange measurements are referred to the thorough general discussion by Sestak et al.[39] Several reviews providing good discussions of photorespiratory gas exchange phenomena have appeared.[8,13,40,41]

A. Post-Illumination Burst (PIB)

Upon darkening a leaf at the end of a period of photosynthesis, a transient burst of CO_2 evolution, the PIB, is observed prior to the establishment of a steady rate of dark respiration. This burst of CO_2 is thought to be related to photorespiration because the magnitude of the PIB responds to the concentration of O_2 and CO_2 during the period of illumination. Low CO_2 concentrations[42] or high O_2 levels[42-45] generally enhance the peak height or the magnitude of the PIB. An exception occurs at very low CO_2 concentrations when the synthesis of the substrate for the photorespiratory pathway is thought to limit the extent of photorespiration.[42]

The PIB is thought to represent a vestige of the flux through the photorespiratory pathway. It is assumed that photosynthesis ceases much more rapidly upon cessation of illumination than photorespiration, and that dark respiration does not occur to a significant extent during the preliminary minutes of darkness.[1,42] Thus, for a brief period, photorespiration can be observed in the absence of photosynthesis and possibly in the absence of dark respiration. However, photorespiratory activity during the PIB is not an accurate reflection of photorespiration under normal photosynthetic conditions. It is probable that flux through the pathway to the CO_2-evolving steps[46,47] occurs at a declining rate as the concentration of photorespiratory intermediates is reduced from steady-state light levels to dark levels.[13] The area under the curve traced by the PIB thus provides a measure of the difference in the amount of carbon in photorespiratory intermediates between light and dark conditions. Doehlert et al.[42] have suggested that with the use of a suitable mathematical transformation, the PIB can provide quantitative estimates of the magnitude of photorespiration. However, because the PIB is a composite measure of several distinct phenomena (i.e., size of the glycine pool, rapidity with which dark respiration resumes, rapidity with which photosynthesis stops), we would not consider it suitable for comparisons between genotypes because of the possible variation in these and other physiological characteristics. Also, transients such as those produced during the PIB can be difficult to quantitate accurately enough for comparative purposes.[13]

It is not necessary to remove all illumination in order to observe PIB phenomena. Following a reduction in light intensity, a photosynthesizing leaf may evolve a burst of CO_2 during a period of readjustment to a lower photosynthetic rate. It has been suggested[48] that the magnitude of this response varies in a manner expected of a photorespiratory indicator, but published evidence to this effect is not yet available. Whatever the case, it cannot be considered a quantitative measure of photorespiration for the reasons already noted with respect to the PIB.

B. CO_2-Evolution into CO_2-Free Air

During illumination in CO_2-free atmospheres containing O_2, the leaves of C_3-species will evolve considerable amounts of CO_2. Under these conditions, CO_2-assimilation is necessarily reduced to the refixation of photorespiratory CO_2. Thus, rates of CO_2 release can be measured directly with little interference from concurrent CO_2-fixation. The rate of CO_2 release into CO_2-free air increases with increases in the O_2 content of the atmosphere,[44,45,49] is dependent upon a preceding period of photosynthesis, increases with temperature, and responds to light intensity.[50,51] These are characteristics which are consistent with a photorespiratory phenomenon.

This method provides a measure of the potential of the leaf for photorespiratory activity, but does not provide a quantitative estimate of the rate of CO_2 release under photosynthetic conditions where the leaf is not entirely dependent on stored carbohydrate as a source of carbon. This measure of photorespiration is probably an underestimate of the rate in air since the supply of substrate for the photorespiratory pathway may be limiting in CO_2-free environments. Much of the substrate under standard atmospheres is thought to arise from recently fixed carbon.[44] In CO_2-free atmospheres, this source would be rapidly depleted and reserve materials such as starch and sucrose would be mobilized to support photorespiration.[52,53] Thus, the rate of CO_2-evolution may, in part, reflect rate-limiting steps in the mobilization of reserve materials rather than the rate which may be obtained during photosynthetic conditions. However, with proper control of plant pretreatment and measuring conditions, CO_2-release into CO_2-free air can provide a relatively reproducible quantitative estimate of photorespiration for comparative purposes.

C. Extrapolation of Apparent Photosynthesis Rates to CO_2 Concentrations of Zero

Extrapolation from the linear portion of the response curve of apparent photosyn-

Table 1
OXYGEN INHIBITION OF APPARENT
PHOTOSYNTHESIS IN INTACT LEAVES[a]

Species	Photosynthesis rate ($mg\ CO_2\ dm^{-1}\ hr^{-1}$)			Percentage inhibition[b]	
	2% O_2	22% O_2	49% O_2	22% O_2	49% O_2
Lolium perenne	22.8	17.1	9.2	25	60
Brassica napus	42.7	31.4	20.0	26	53
Spinacea oleracea	33.6	23.6	14.2	30	58

[a] Measurements were made at 25°C and 54% relative humidity at a CO_2 concentration of 350 $\mu l\ l^{-1}$ with a light intensity of 1150 μ Einsteins PAR m^{-2} sec^{-1}.
[b] Relative to the rate in 2% O_2.

thesis as a function of CO_2 concentration to a CO_2 concentration of zero provides a measure of photorespiration.[40,54] The intercept value is proportional to the O_2 concentration used for CO_2-fixation measurements. At 2% O_2, little or no photorespiratory activity is detected by this method. Both light intensity and the flow rate used when determining rates of CO_2 assimilation can influence this measurement of photorespiration.[55] An underlying assumption of this method is that the rate of photorespiration is not influenced by CO_2 concentration over the range needed for extrapolation. There is no theoretical basis for this assumption. The method is not well suited for multiple comparisons because of the necessity of making many measurements on each plant.

D. CO_2-Compensation Concentration (Γ)

If a C_3 leaf is illuminated in a closed gas exchange system containing 21% O_2, it will reduce the level of CO_2 in the atmosphere to a constant level of about 40 to 60 μl CO_2 l^{-1}. This concentration is referred to as the CO_2 compensation point (Γ). It is characteristic of plants with measurable compensation points that they will either release CO_2 into CO_2-free atmospheres or reduce the CO_2 concentration in a closed container until Γ is reached. This value represents the CO_2 concentration at which CO_2 assimilation and evolution are balanced so that no net uptake or evolution occurs.[6] Increases in either the O_2 concentration or temperature lead to increased values for Γ.[45,56] This measure of photorespiration is a relatively good indicator of the photorespiratory potential of a leaf. It provides a generally reproducible measure which can be utilized for comparisons among treatments or genotypes.[56-58] However, it cannot be used to determine flux through the photorespiratory pathway. Also, it can be technically difficult to construct a sufficiently tight gas exchange system because CO_2 is very permeable to many of the substances commonly used for tubing.[59] Even small leaks can result in significant overestimates of Γ. This measure has been used as a rapid method to classify C_3 and C_4 species.[57] A report that the value may undergo a seasonal variation remains anomalous.[56]

E. The Magnitude of O_2-Inhibition of Photosynthesis

The inhibitory effect of O_2 on photosynthesis has been termed the "Warburg effect". The magnitude of the effect is frequently expressed as the percentage inhibition of apparent photosynthesis in an O_2 enriched atmosphere as compared to the rate observed in 2% O_2 (Table 1). As noted earlier, (Section II), one component of the Warburg effect is photorespiration. Factors which enhance photorespiration, such as low CO_2 levels or high O_2 levels, exacerbate the amount of O_2-inhibition of photosynthesis.[60,61] The Warburg effect is a composite of all the effects of O_2 on photosynthesis.

These include direct inhibition of RuBP carboxylation, enhanced photorespiration,[6,45] and Mehler-like reactions.[62] Thus, the absolute difference in CO_2-fixation rate between 2% and 21% O_2 cannot be used as a quantitative estimate of photorespiratory fluxes but is a good indicator of the combined effects of RuBP oxygenase activity on net photosynthesis. The measure also reflects photosynthetic and photorespiratory capacity in steady-state conditions.

F. Release of $^{14}CO_2$ into CO_2-Free Air

Zelitch[63] has advocated a method in which the evolution of $^{14}CO_2$ by leaf discs is used as a measure of photorespiratory activity. The discs are first allowed to photoassimilate $^{14}CO_2$ for about 45 min; then they are flushed with CO_2-free air in the light, followed by a flushing period in the dark. Comparisons among plants or treatments are made after values for $^{14}CO_2$ evolution in the light are standardized against similar measures made in the dark.

Because the specific activity of the evolved $^{14}CO_2$ may vary over the course of the measuring period,[52] measures of $^{14}CO_2$ evolved are not indicative of total photorespiratory CO_2 evolution. The method has been criticized in some detail.[64]

G. Dual Isotope Methods

Ludwig and Canvin[52] described a method which permits simultaneous measurements of CO_2 assimilation and photorespiration in an open gas exchange system under steady-state conditions in a wide variety of atmospheres. This is accomplished by first bringing a leaf to steady-state photosynthesis in $^{12}CO_2$. The gas is then quickly switched to one containing the same level of CO_2 and O_2 but supplemented with $^{14}CO_2$ of known specific activity. Initially, a very rapid uptake of $^{14}CO_2$ is observed which represents true photosynthesis in the absence of photorespiratory CO_2 release. This measure must be made in less than 15 sec,[8,65] since after this time $^{14}CO_2$ begins to be evolved from recently fixed carbon and diminishes the observed rate of uptake. The difference between the maximal rate of $^{14}CO_2$ uptake and the steady-state rate of $^{12}CO_2$ uptake is considered to be a measure of the rate of photorespiratory CO_2 release.

This method assumes that $^{14}CO_2$ is a suitable tracer for $^{12}CO_2$ and that little or no discrimination between the isotopes occurs. This assumption seems valid because the two isotopes compete with one another.[65] The method also assumes that the specific activity of $^{14}CO_2$ at the site of fixation is not significantly different from that supplied in the gas stream. An implicit extension of this assumption is that the amount of photorespired CO_2 which is refixed without being released to the atmosphere is not substantial. The validity of the second assumption is difficult to assess from gas exchange measurements alone, and the degree to which refixation of photorespired CO_2 does occur results in a corresponding underestimate of true photosynthesis and photorespiration.[66]

An apparent anomaly that has arisen in the use of the dual carbon isotope method is that photorespiration appears to be relatively insensitive to the CO_2 concentration of the gas stream by this method.[65] Ku and Edwards[67,68] have suggested that the refixation of photorespired CO_2 may be proportionately greater at low CO_2 concentrations. This could result in a CO_2-dependent bias which would lead to a greater underestimate of photorespiration at low CO_2 concentrations.[66] Alternatively, in consideration of the in vitro kinetic properties of RubisCO, only a slight reduction in photorespiratory CO_2 evolution would be expected over the range of CO_2 concentrations tested. The dual isotope method may simply not be sensitive enough to detect the corresponding changes in photorespiration.

$^{18}O_2$ uptake and $^{16}O_2$ evolution can also be used to simultaneously monitor photorespiration and photosynthesis in standard atmospheric conditions.[69,70] Measurements are performed using $^{18}O_2$ of high specific activity in a closed gas exchange system. Changes in the amount of $^{18}O_2$ and $^{16}O_2$ are monitored with a mass spectrometer. Estimates of photorespiration obtained by this method are complimentary to those obtained by monitoring CO_2 fluxes, but are not biased by CO_2-refixation.[41] Uptake of $^{18}O_2$ responds to changes in atmospheric levels of CO_2 and O_2, decreasing as CO_2 is increased,[70] and increasing with increases in O_2.[62,70,71] However, anomalous observations at low CO_2 concentrations[70] or high light intensities[62] are not readily explained in the context of photorespiration. Mehler-type O_2 consumption has been invoked to account for $^{18}O_2$ uptake patterns under conditions in which CO_2-fixation is limiting.[72,73]

The use of oxygen isotopes requires that there be no discrimination between the two isotopes, and that no significant refixation of evolved $^{16}O_2$ occurs. In suspensions of algal cells[74] or chloroplasts,[71] little refixation of endogenous O_2 occurs and $^{18}O_2$ uptake reflects total uptake quite precisely. However, it has been suggested[71] that refixation of endogenous O_2 may be more pronounced with leaves than with isolated chloroplasts. Although the O_2-based dual isotope method has some potential advantages over the carbon isotope ratio method, it has been less widely utilized because of the expense of $^{18}O_2$ and the detection systems, and the need to use closed gas exchange systems.

H. Evaluation of Various Methods

All measurements of the rate of photorespiration based upon gas exchange parameters are unavoidably indirect. Gas exchange measures are net or composite values of all the processes which contribute to CO_2 and O_2 exchanges between the cellular environment and the external atmosphere.[8] These include factors not directly related to the photorespiratory pathway such as the relative solubility of O_2 and CO_2 in liquid phases.[67,68] Leaf architecture undoubtedly affects diffusive resistance to gaseous exchange with the environment[75] and, therefore, the amount of refixation which occurs. Photorespiration can be overestimated to the extent that nonphotorespiratory processes evolve CO_2 or consume O_2. At present, it is generally assumed that processes such as dark respiration and Mehler-type reactions are not major components of the observed gas exchanges under standard photosynthetic conditions. However, under more extreme conditions, such as superoptimal temperature or light intensity, very high O_2 or very low CO_2 concentrations, these processes may make a relatively more important contribution to gas exchange measures.

In order to monitor photorespiration by gaseous exchange, it is advisable to use methods which directly measure flux through the photorespiratory pathway, and to minimize the effect of diffusive resistance and ancilliary CO_2-evolving or O_2-consuming processes. The use of isotopic variants of CO_2 or O_2 to monitor photorespiration during active photosynthesis is probably the most sound method available at present for assessing photorespiration. With a more modest investment in equipment, photorespiratory potential can be assessed by measuring the Warburg effect on apparent photosynthesis, the CO_2-compensation concentration, or the rate of CO_2 release into CO_2-free air. Because these latter measures are indirect, any genotypic variation should be confirmed by more than one method.

A final point is that since the Calvin cycle provides the substrate for photorespiration, the amount of photorespiration will necessarily be proportional to the amount of Calvin cycle activity. Treatments which inhibit photosynthesis will also inhibit photorespiration.[7,76] Thus, variability in the amount of photorespiration is of dubious significance if it is correlated with a similar variation in the rate of photosynthesis.

V. THE BIOCHEMISTRY OF PHOTORESPIRATION

A. Phosphoglycolate

The primary reaction of photorespiration is the oxygenation of RuBP to produce one molecule each of 3-PGA and P-glycolate. P-glycolate is a potent inhibitor of the Calvin cycle enzyme triose phosphate isomerase.[77,78] The concentration of this compound in the chloroplasts is, therefore, maintained at very low levels by the presence of a very active and specific phosphatase, which is found in both C_3 and C_4 plants.[79,80] The enzyme exhibits a high degree of substrate specificity[81] which is obviously required to prevent wasteful hydrolysis of other Calvin cycle intermediates. No specific inhibitor of the activity is known, although glycidol-phosphate (2,3-epoxy-propanol phosphate) has been used for mechanistic studies.[81]

Evidence for the importance of this enzyme, and for the massive flux of photorespiratory carbon through P-glycolate, was provided by the isolation of a mutant of *Arabidopsis thaliana* (L.) lacking this activity.[82] Under conditions which permitted RuBP oxygenation, the mutant accumulated P-glycolate, and photosynthesis was severely impaired by the resulting inhibition of triose phosphate isomerase. However, in atmospheric conditions which prevented RuBP oxygenation (i.e., 1% CO_2 in air), the mutant exhibited normal growth indicating that the enzyme has no other necessary function. Two physically distinguishable forms of the enzyme have been observed in some species,[83] but the physiological significance of this observation is uncertain.

B. Glycolate

The glycolate which results from P-glycolate hydrolysis is not further metabolized within the chloroplast.[71,84,85] Earlier studies indicating synthesis of glyoxlate, glycine, or serine from glycolate by isolated chloroplasts are considered to have been due to contamination of chloroplast preparations with other organelles.[8] It is not certain how glycolate is exported from the chloroplast. Recent studies of glycolate transport by isolated chloroplasts have not provided any evidence for a specific carrier,[86-88] but the observed rate of diffusion appears too slow to account for measured rates of glycolate production in vivo.[88] Since the subsequent metabolism of glycolate takes place in peroxisomes, it is possible that there exists a fragile mechanism, easily disrupted by the usual organelle isolation techniques, which ensures the directed flow of glycolate from chloroplasts to peroxisomes. Improved methods for isolating peroxisomes may permit a novel approach to this problem.[89,90]

The only known metabolic fate for glycolate in leaf cells is oxidation to glyoxylate. This reaction is irreversibly catalyzed in higher plants by the peroxisomal flavoprotein, glycolate oxidase.[91,92] During glycolate oxidation, enzyme-bound FMN is reduced and then reoxidized by molecular O_2 to produce H_2O_2. The H_2O_2 is decomposed to H_2O and $1/2$ O_2 by catalase which is also present solely in peroxisomes.[3]

Glycolate oxidase can also oxidize L-lactate but not D-lactate, and is insensitive to cyanide. These features distinguish the enzyme found in all higher plants, including mosses, ferns, liverworts, and freshwater angiosperms, from a glycolate dehydrogenase found in some species of freshwater algae.[93] The latter enzyme does not use O_2 as a terminal electron acceptor, catalyzes glycolate-dependent DCPIP-reduction, oxidizes D-lactate, and is cyanide sensitive. Also, in contrast to the higher plant enzyme, glycolate dehydrogenase is not located in a microbody of the peroxisomal type but appears to be associated with the mitochondrion.[94] This raises the possibility that algae conserve energy associated with glycolate oxidation by coupling it to electron transport. There has been one report of an NADH-linked glycolate dehydrogenase in higher plants,[95] but the work appears to have been discontinued.

A number of glycolate oxidase inhibitors have been identified and used for metabolic studies. These include aldehyde bisulfite addition compounds,[96] of which α-hydroxypyridine methane sulfonic acid (α-HPMS) is the most widely used. This compound also inhibits phosphoenol pyruvate (PEP) carboxylase, NAD-malate dehydrogenase,[97] and probably aminotransferases,[3] reduces photosynthesis,[98,99] and is strongly herbicidal.[96]

A theoretically more specific inhibitor is the suicide analog hydroxybutynoic acid (HBA). This compound irreversibly inhibits glycolate oxidase by reaction of the acetylene group with enzyme-bound FMN.[92] The compound has been used for measuring glycolate accumulation,[82,100] but also inhibits photosynthesis in normal atmospheric conditions.[101,102]

C. Glyoxylate

Glyoxylate, the product of the glycolate oxidase reaction, is a very reactive compound and may undergo a variety of theoretically possible metabolic fates. The degree to which these alternate possibilities are realized in vivo may exert a strong effect on the stoichiometry of photorespiration.

The predominant route of metabolism appears to be transamination of glyoxylate to glycine. Peroxisomes contain two very active glyoxylate aminotransferases which utilize serine, alanine, glutamate, and, with lower efficiency, several other amino acids as substrates.[103] The glutamate:glyoxylate aminotransferase has been purified and shown to also utilize alanine but not serine as amino donor.[104] Purified serine:glyoxylate (hydroxypyruvate) aminotransferase also utilizes alanine but not glutamate.[105] Both the glutamate:glyoxylate aminotransferase reaction and the serine:glyoxylate aminotransferase reaction are essentially irreversible.[103] This property is unusual since most aminotransferase reactions are freely reversible.[106] The competition between serine and glutamate as amino donors for glyoxylate amination has been examined in isolated peroxisomes.[107] Serine was preferred amino donor, as would be predicted by the scheme in Figure 1, in which uninterrupted flux of carbon through the photorespiratory pathway depends upon quantitative deamination of one serine for each two glycines formed. In this way, the requirement for amino donors for quantitative glyoxylate transamination is minimized by recycling the amino groups between glycine and serine.

The importance of the serine:glyoxylate aminotransferase as an essential component of photorespiration was demonstrated by the isolation of several mutants of *Arabidopsis* which were deficient in this enzyme activity.[108] The mutants showed normal growth under conditions where photorespiration was prevented by high levels of atmospheric CO_2, but had greatly reduced photosynthesis and were inviable in air. Serine accumulated to high concentrations in the mutants, indicating a very limited range of quantitatively significant metabolic fates for photorespiratory serine.

Glyoxylate is not only a product, but may also be a substrate of glycolate oxidase. The product of this reaction is oxalate, which is present in substantial quantities in the leaves of some species. Isolated peroxisomes will produce low amounts of oxalate from exogenous glycolate, but glycine is by far the major product if an amino donor is available.[109,110] Photorespiration does not appear to play a significant role in oxalate synthesis in those plants which accumulate significant quantities of this compound.[111]

Glyoxylate reacts spontaneously with H_2O_2 to produce CO_2 and formate, and catalase may catalyze peroxidative decarboxylation of glyoxylate.[112] Since glyoxylate and peroxide are produced stoichiometrically and, presumably, in intimate proximity by the action of glycolate oxidase on glycolate, it has been suggested that this reaction could account for a substantial proportion of the CO_2 evolved in photorespiration.[113-115] However, catalase represents about 15% of the total protein in peroxisomes,[3] and the specific activity of the enzyme is 100- to 1000-fold greater than that of the other peroxisomal enzymes. Thus, it seems likely that H_2O_2 exists only transiently and would not be available to react with glyoxylate under normal circumstances. Several authors have followed the fate of ^{14}C-glycolate provided to isolated peroxisomes and under conditions of moderate temperature and availability of amino donors have not observed significant amounts of CO_2 release.[109,116] Advocates of the scheme[112,114,115,117] have shown that isolated peroxisomes can be induced to release $^{14}CO_2$ from ^{14}C-glycolate. However, the conditions necessary to achieve this are relatively harsh, and produce relatively low amounts of $^{14}CO_2$ in vitro. Although it is difficult or impossible to extrapolate to the in vivo situation, it has been suggested that as much as 15 to 20% of photorespired CO_2 may arise in this reaction, particularly at high temperature. Formate produced in this reaction could, in principle, be oxidized to CO_2 by a low-activity mitochondrial NAD-linked formate dehydrogenase[118] or incorporated into the β-carbon of serine by a relatively more active formyl tetrahydrofolate synthase in leaves.[119] The amount of the formate dehydrogenase activity varies between species, being absent in soybean mitochondria but very active in spinach.[120]

Some evidence that glyoxylate decarboxylation may occur in vivo under extreme conditions has been obtained with a mutant of *Arabidopsis* which is unable to metabolize photorespiratory glycine because of a defect in serine transhydroxymethylase activity.[47] Because of this defect, the mutant rapidly became depleted of amino donors for glyoxylate amination following exposure to photorespiratory conditions. In these circumstances, glyoxylate did not accumulate but was apparently lost as CO_2. This efflux of CO_2 was abolished by the provision of exogenous NH_3 which permitted transamination to continue. This observation strongly suggests that, in the presence of amino donors, the glycine decarboxylase reaction is the only source of photorespiratory CO_2, but that under conditions which limit the availability of amino donors, glyoxylate decarboxylation may occur. Although it is not known if this situation ever prevails under field conditions, it could provide a plausible explanation for the observation that nitrogen-deficient bean and cotton had greater O_2-inhibition of photosynthesis than the adequately nourished counterpart.[121]

The remaining, frequently noted possible fate of glyoxylate is transport to the chloroplast and reduction to glycolate by chloroplast NADPH-glyoxylate reductase.[122] This, of course, leads to futile consumption of NADPH since the only metabolic fate of glycolate is reoxidation to glyoxylate. The motivation for this scheme is the presence in the chloroplast of an NADPH-glyoxylate reductase. However, the activity of this enzyme is low (*ca.* 5%) relative to the peroxisomal enzymes,[123] and the intracellular concentration of glyoxylate is probably too low to permit such a shuttle.[8] Although there is no evidence for such a shuttle, and its existence does not appeal to common sense, it has been invoked to explain the reportedly stimulatory effects of glycidate on photosynthesis.[124] Also, anomalous tritium enrichment in glycolate following dual labeling with ^3H_2O and $^{14}CO_2$ has been interpreted as being consistent with the operation of a shuttle.[125] However, Lorimer[293] has suggested that the ability of glycolate oxidase to catalyze ^3H-exchange may be responsible for the unusual ^3H-enrichment pattern. In the absence of other evidence, there is little reason to assume that a shuttle operates.[8]

D. Glycine and Serine

Glycine metabolism appears to be associated primarily with the mitochondrion. The mechanism by which it moves from the peroxisome to the mitochondrion is not clear but there is evidence suggesting the presence of a specific carrier in the mitochondrial membrane.[126-129] Day and Wiskich[130] were unable to find evidence for such a carrier, possibly because they did not preincubate mitochondrial preparations with the inhibitor (mersalyl) used to define the carrier in other studies.

The details of the mitochondrial reactions involving glycine have been difficult to study directly because glycine metabolism is abruptly abolished when leaf mitochondria are ruptured.[46] Early studies with relatively crude preparations demonstrated that leaf mitochondria catalyzed the synthesis of 1 mol each of serine, CO_2, and NH_3 from 2 mol of glycine,[3] and that glycine cleavage could be coupled to ATP synthesis.[131,132] Subsequent studies[133-136] confirmed that glycine oxidation is linked to electron trans-

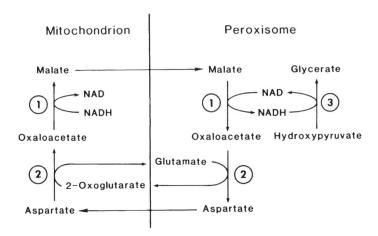

FIGURE 2.   A plausible scheme for indirect transfer of reducing equivalents from the mitochondrion to the peroxisome. Because stoichiometric amounts of NADH and hydroxypyruvate result from mitochondrial glycine decarboxylation, this scheme would provide a means for ensuring the uninterrupted forward flux of photorespiratory carbon. Specific transport systems mediate the transfer of the dicarboxylates across mitochondrial membranes. The relevant enzymes are (1) malate dehydrogenase; (2) glutamate:oxaloacetate aminotransferase; (3) hydroxypyruvate reductase.

port and is coupled to three phosphorylation sites, suggesting the involvement of NADH. The stoichiometry of the overall reaction is two CO_2 released per O_2 consumed, which suggests the generation of one NADH produced per glycine decarboxylated.[137]

Although the coupling to the respiratory chain proceeds well in vitro, it is improbable in vivo because the cytoplasmic ATP:ADP ratio is unfavorable during illumination of leaves.[138] It has therefore been suggested[136,139] that reoxidation of mitochondrial NADH may be accomplished by indirect shuttling of reducing equivalents from the mitochondrion via a malate shuttle as illustrated in Figure 2. Plant mitochondria are capable of transporting compounds such as malate and aspartate which are thought to be involved in indirect transfer of reducing equivalents, and have the necessary matrix enzymes (glutamate: oxaloacetate transaminase and malate dehydrogenase). In support of the shuttle hypothesis, it is observed that glycine-dependent O_2 consumption by isolated mitochondria is suppressed by exogenous 2-oxoglutarate plus aspartate.[139] The existence of an NADH reoxidizing system is of some importance because of the likelihood that the glycine cleavage reaction could otherwise be limited in vivo by NAD^+ availability.[140]

Details of the glycine cleavage reaction in plants are unknown. It has generally been assumed that the mechanism is similar to, or identical with, glycine-cleavage reactions in bacteria and vertebrates.[141-143] In these organisms, where it has been possible to solubilize and purify the enzymes involved, the overall reaction has been subdivided into two sequential reactions described by the equations:

$$\text{Glycine} + H_4 \text{ folate} + \text{NAD} \rightarrow CO_2 + NH_3 + CH_2H_4 \text{ folate} + \text{NADH} \qquad (4)$$

$$\text{Glycine} + CH_2H_4 \text{ folate} \rightarrow \text{serine} + H_4 \text{ folate} \qquad (5)$$

The reaction in Equation 5 is catalyzed by the enzyme serine transhydroxymethylase, which is present in leaves as two isoenzymes, one of which is located in the mitochondrial matrix.[47,144] The reaction catalyzed by this enzyme is reversible but appears to proceed primarily in a forward direction in vivo since exogenously provided serine does not rapidly label the glycine pool of leaves in the light.[145]

In vertebrates, the enzyme complex that catalyzes Equation 4 (designated glycine synthase) consists of four protein components that can be individually purified and reconstituted to form a catalytically active complex.[141-143] One of these four proteins, glycine decarboxylase, contains pyridoxal phosphate, and catalyzes the cleavage of glycine to CO_2 and methylamine without the addition of other cofactors.[141] This protein also catalyzes the exchange of bicarbonate with the carboxyl group of glycine, a reaction also observed with intact plant mitochondria.[128,136,146,147] Both reactions are greatly stimulated by the addition of a lipoic acid-containing protein (aminomethyl carrier protein)[142] which acts as an electron acceptor and as a carrier of the methylamine resulting from glycine decarboxylation. The presence of the other two components of the complex is required for the transfer of electrons to NAD^+, the release of NH_3, and the transfer of the methylene group to tetrahydrofolate.

Mutants of *Arabidopsis* lacking the mitochondrial isoenzyme of serine transhydroxymethylase,[47] and another class deficient in the glycine decarboxylase reaction[128] have been isolated. Both classes of mutants lack photorespiration under conditions which ensure an adequate supply of amino donors for glyoxylate transamination, indicating that the glycine decarboxylase reaction is the sole site of photorespiratory CO_2 release under normal conditions. The normal growth of these mutants in atmospheres which prevent the oxygenation of RuBP indicates that these enzymatic activities are not required for any function other than photorespiratory metabolism.

Several inhibitors of the glycine cleavage reaction have been identified. Isonicotinic acid hydrazide (INH), the vitamin B6 antagonist, blocks the reaction,[113] possibly by interacting with the pyridoxal phosphate-requiring component of the glycine synthase complex. INH applied to leaves inhibits photosynthesis.[101] Glycine hydroxamate (GH) has also been reported to exhibit reversible inhibition[148] but also reacts spontaneously with glyoxylate and is slowly metabolized. Aminoacetonitrile (AAN) is a potent and apparently specific inhibitor of glycine decarboxylation.[149] The effect of AAN on photosynthesis by soybean leaf discs[150] is very similar to the effects of the mutations in *Arabidopsis* which block glycine decarboxylation. The precise mode of action of these compounds is not known. It has been reported that INH, GH, and AAN inhibited both glycine-bicarbonate exchange and serine transhydroxymethylase activity,[147] possibly by affecting the transport of glycine into mitochondria rather than affecting enzyme activity per se.[129]

The glycine decarboxylase activity is apparently present only in photosynthetic tissue, and is not found in mitochondria from roots, stalk, or veins of spinach.[151] The development of the activity is apparently controlled by light since dark grown seedlings of sunflower had very low levels of activity, but 5 min per day of illumination was sufficient to induce high levels of activity.[152] No activity was found in potato tuber mitochondria, etiolated mung bean hypocotyls, or mitochondria from maize leaves.[153] These observations support the notion that the activity is required only for metabolism of photorespiratory glycine.[128]

E. Terminal Steps of the Pathway

Serine formed in the mitochondrial serine transhydroxymethylase reaction returns to the peroxisome where it serves as an amino donor for glyoxylate transamination. Serine is the preferred amino donor in isolated peroxisomes,[107] thereby ensuring flow through the pathway. Early experiments[154] demonstrated a light requirement for serine metabolism which is now recognized to be due to the fact that serine deamination depends on the availability of glyoxylate as an amino acceptor. Since glyoxylate is not formed in the dark, the metabolism of serine is slow in leaves supplied with ^{14}C-serine in the dark. A mutant lacking serine: glyoxylate aminotransferase accumulated serine in large amounts, and the pool of accumulated serine turned over extremely slowly,[108] indicating that there is no other quantitatively significant route of serine metabolism in leaf cells.

The product of the serine:glyoxylate aminotransferase reaction is hydroxypyruvate. This compound, like glyoxylate, is relatively unstable and is subject to rapid decarboxylation when provided to isolated peroxisomes[110] or detached leaves.[155] The conversion of hydroxypyruvate to glycerate is catalyzed by the only peroxisomal reductase, NADH-hydroxypyruvate reductase,[123] which was originally described as an NADH-glyoxylate reductase. Although the reverse reaction (D-glycerate dehydrogenase) can be shown with isolated peroxisomes, it is not considered important because of an unfavorable pH optimum.

Serine has long been known to be a precursor of sucrose in C_3 leaves. The presumed route is via hydroxypyruvate and glycerate to PGA. There has been, however, some uncertainty as to whether glycerate carbon must first enter the chloroplast or whether there is an entirely cytoplasmic route to sucrose. The localization of the key enzyme, glycerate kinase, which catalyzes the ATP-dependent phosphorylation of D-glycerate to PGA, is a critical piece of evidence in this particular puzzle. An early report[156] suggested that some glycerate kinase was cytoplasmic, leading to speculation that photorespiratory carbon could be converted to sucrose without reentering the chloroplast.[157] This contrasts with the situation in which the kinase is entirely within chloroplasts. In this case, conversion of glycerate to PGA in the chloroplast would necessarily result in the carbon of the glycolate pathway entering a common pool of triose phosphate. Usuda and Edwards[158] showed that glycerate kinase is located entirely in chloroplasts in C_3 species. Interestingly, the enzyme appears to be restricted to mesophyll cells of C_4 species implying that, if C_4 plants do in fact exhibit some flow through the photorespiratory pathway,[13] the glycerate formed must be transported from the bundle sheath to the mesophyll for conversion to PGA.

Recently, Robinson[159] has shown light stimulation of glycerate transport into chloroplasts and provided kinetic evidence for a specific glycerate transporter in chloroplast envelopes.[160] The existence of this transporter, the requisite enzymes, and the results of early tracer studies[154,161] indicating flow of label from serine to sugars via glycerate support the formulation of the photorespiratory pathway shown in Figure 1.

F. Other Mechanisms of Serine and Glycine Synthesis

It has been suggested[162] that photosynthetic carbon could flow from PGA to serine by a reversal of the terminal steps of the photorespiratory pathway. The primary motivation for this hypothesis appears to have been the report of a chloroplast enzyme which specifically hydrolyzes PGA to glycerate.[80,163] However, the existence of this phosphatase might be considered somewhat unusual because of the necessity of imposing unusual and stringent regulation on the enzyme to prevent futile hydrolysis of Calvin cycle intermediates. Recently, the localization of this enzyme in chloroplasts has been challenged by a report[164] suggesting that the apparent localization was due to contamination of broken chloroplasts by acid phosphatase. Thus, the primary motivation for considering reverse flow is in doubt. In addition, the hydroxypyruvate reductase is not well suited as a glycerate dehydrogenase under physiological conditions,[3] and there is no report of a peroxisomal hydroxypyruvate transaminase which could convert hydroxypyruvate to serine. Finally, Mahon et al.[165] concluded from comparisons of the specific activity of glycine and PGA in 1% CO_2 that their results were inconsistent with formation of glycine from serine via reverse flow of the photorespiratory pathway. The concept of reverse flow of carbon from PGA to serine appears to be largely unsubstantiated speculation.

An alternate possibility is that glycine and serine could be produced from PGA via phosphohydroxypyruvate and phosphoserine. The necessary enzymes have been reported to be present at low levels in spinach chloroplasts[166] and a chloroplast isoenzyme of serine transhydroxymethylase has been reported.[146] In mutants of *Arabidopsis* unable to metabolize photorespiratory glycine to serine, less than 1% of the $^{14}CO_2$ incorporated is found in serine during steady-state photosynthesis.[155] This experiment supports the notion of a route of serine synthesis not involving glycine, but clearly indicates that it is quantitatively minor by comparison with the magnitude of flux through the forward direction of the photorespiratory pathway.

G. Photorespiratory NH₃ Cycle

During the mitochondrial decarboxylation of glycine, NH_3 and CO_2 are released in stoichiometric amounts.[167] There is abundant evidence that the NH_3 is rapidly refixed into glutamine by ATP-dependent glutamine synthetase (GS) which exists in two isoforms distributed in the cytosol and the chloroplast.[168,169] There is no convincing means of distinguishing which of these enzymes is primarily involved. Glutamine is utilized in the chloroplast where one molecule of glutamine and 2-oxoglutarate are converted to two molecules of glutamate by the ferredoxin-dependent glutamate synthase (GOGAT). The chloroplast envelope has a specific carrier which permits glutamine to penetrate.[170] Another carrier, frequently designated the dicarboxylate transporter, facilitates the counter exchange of compounds such as malate, aspartate, 2-oxoglutarate, and glutamate across the envelope.[86,170] Thus, the glutamate, which is required as a primary NH_3 acceptor in the GS reaction, and as an amino donor for glyoxylate metabolism is regenerated by GOGAT, thereby completing the photorespiratory ammonia cycle[167] (Figure 1).

Convincing evidence for the importance of the GS/GOGAT cycle and the dicarboxylate transporter in photorespiration has been obtained by the isolation of mutants lacking these functions.[170,171] The normal growth of these mutants under conditions which prevent flow of carbon into the photorespiratory pathway suggests the primary function of these activities is photorespiratory ammonia recycling.

The refixation of NH_3 by the GS/GOGAT pathway requires one ATP and one reduced ferredoxin (equivalent to one NADPH). Although this increases the demand for ATP and NADPH, the increase is relatively insignificant compared with the base level of ATP and NADPH consumption associated with the regeneration of RuBP for PGA during CO_2 assimilation.[11]

H. Source of Photorespiratory CO₂

As noted in the foregoing discussion, both glycine and glyoxylate have been proposed as the immediate precursors of photorespiratory CO_2. The resolution of this uncertainty is of some importance because these alternatives could dramatically alter the stoichiometry of the photorespiratory pathway. If glycine decarboxylation is the sole mechanism responsible for photorespiratory CO_2 release, then one of four carbon atoms entering the photorespiratory pathway will be lost as CO_2. If, on the other hand, glyoxylate is oxidized directly and the resulting formate is also oxidized,[112,120] all the carbon would be lost — a fourfold difference.

Attempts to resolve this issue by comparing the specific activity of the carbon atoms in glycolate, glycine, or glyoxylate with that of evolved CO_2 following $^{14}CO_2$ incorporation[172,173] have substantiated the involvement of these compounds but have not discriminated among them. Inhibition of the glycine decarboxylation reaction by chemical inhibitors has also failed to provide unequivocal evidence for a single site of

CO_2 release although photorespiration appears to be greatly reduced by such treatment.[100,101,140,148] The problem with such experiments is that it is not possible to know with certainty that the desired chemical inhibition is absolutely effective. Also, it is possible that by blocking one reaction, another reaction, which is not normally employed, may come into play to relieve stress on the system.

Evidence that glycine decarboxylation is the sole mechanism of photorespiratory CO_2 release under normal conditions was obtained with mutants of *Arabidopsis* unable to decarboxylate glycine.[47,128] When leaves of the mutants are illuminated in atmospheres which stimulate flow of carbon into the photorespiratory pathway, glycine accumulates in large amounts so that mesophyll cells become depleted of amino donors for glyoxylate transamination. As a result, glyoxylate is decarboxylated, leading to photorespiratory CO_2 release. Thus, glyoxylate decarboxylation can occur under extreme circumstances. However, when the leaves were provided with a source of ammonia so that an adequate supply of amino donors for glyoxylate transamination was ensured, no photorespiration was observed.[47] The implication is that glyoxylate decarboxylation occurs only under conditions of amino group depletion. These experiments also revealed a weakness in previous chemical inhibition studies in which depletion of amino donors by inhibitor-induced glycine accumulation had not been considered. Oliver[174] reexamined the effects of inhibitors in the presence of exogenously provided amino donors and found, in contrast to an earlier study, that most of the CO_2 released from metabolites of glycolate could be accounted for by glycine decarboxylation. He estimated that only about 2.5% of photorespiratory CO_2 would arise from glyoxylate decarboxylation. Oliver[174] also measured glyoxylate concentrations in nitrogen-starved plants without noticing any pronounced differences, but did not measure the effect of this treatment on photorespiratory CO_2 evolution directly. It seems, therefore, that plants may decarboxylate glyoxylate directly under stress conditions, but that under nonstress conditions this does not occur.

I. Magnitude of Flux through the Photorespiratory Pathway

Both the enhancement of dry matter production under conditions which suppress photorespiration and the lack of apparent utility of the pathway suggest photorespiration could be profitably eliminated. Inherent in the argument is the concept that flux through the photorespiratory pathway represents a significant portion of plant carbon metabolism. Zelitch[175] has compiled a list of measurements which indicate that photorespiratory CO_2 release is in the range of 14 to 75% of the rate of apparent photosynthesis. The broadness of the range reflects the difficulty of obtaining an accurate measurement.

Flux of intermediates through the photorespiratory pathway can be inferred from measurements of either CO_2 release or O_2 uptake using dual isotope methods. The relative magnitude of photorespiration (μmol carbon entering the photorespiratory pathway per μmol apparent (net) carbon fixed) ranges from 0.67 to 0.85 by the $^{14}CO_2/^{12}CO_2$ method[165,176] and from 1.33 to 1.36 using the $^{18}O_2/^{16}O_2$ method.[62,177] The discrepancy between the two measures may reflect, in part, error introduced by refixation of photorespired CO_2 with the $^{14}CO_2/^{12}CO_2$ method, and by the contribution of nonphotorespiratory mechanisms to $^{18}O_2$ uptake with the latter method.

The flux of intermediates through the photorespiratory pathway may be estimated by measuring the change in specific activity of a photorespiratory intermediate, commonly glycine, upon switching the atmosphere from one containing $^{14}CO_2$ (or $^{16}O_2$) to one containing $^{12}CO_2$ (or $^{18}O_2$) or vice versa. The decay of the specific activity of glycine reflects flux out of the glycine pool. Rates of photorespiration inferred from such measurements are 71 μg carbon dm^{-1} min^{-1} (about 24 μmol CO_2 evolved mg chl^{-1} hr^{-1}) in sunflower,[165] and 4.5 mg carbon as glycine dm^{-1} hr^{-1} (about 31 μmol CO_2 evolved mg chl^{-1} hr^{-1}) in wheat.[178] Berry et al.[71] estimated that the flux at the CO_2 compensation concentration (45 μl CO_2 l^{-1}) and 20% O_2 was 78 μmol glycine mg chl^{-1} hr^{-1} (about 39 μmol CO_2 evolved mg chl^{-1} hr^{-1}) in spinach monitoring $^{18}O_2$ in glycine. The first two measures of photorespiration were made under approximately standard conditions and the calculated ratio of micromole carbon entering the photorespiratory pathway per micromole apparent carbon fixed is 1.2, a value slightly less than that obtained by measuring $^{18}O_2$ uptake with a mass spectrometer.

Estimates of photorespiratory fluxes may also be obtained by using models based on the in vitro kinetic properties of RubisCO. One model based solely on the properties of RubisCO predicted photorespiratory CO_2 evolution in standard atmospheric conditions would be 14% of apparent CO_2 fixation (about 0.56 μmol carbon entering the photorespiratory pathway per micromole apparent carbon fixed).[6,7] A more elaborate model suggests photorespiratory CO_2 release is 21% of net CO_2 fixation (about 0.86 μmol carbon entering the photorespiratory pathway per micromole apparent carbon fixed) (from Figure 7 of Farquhar et al.[18]).

There have been several attempts to obtain quantitative estimates of flux by measuring the amount of photorespiratory intermediates which accumulates in the presence of a particular enzyme inhibitor. For instance, glycolate accumulation in leaf material treated with the glycolate oxidase inhibitors HBA,[100,179,180] or α-HPMS,[181] and glycine accumulation in the presence of INH,[100] have been used as estimates of photorespiratory flux. These values for photorespiration are generally lower than those obtained using other methods, and are flawed by uncertainties in equating mass with label.[9,10] Also, all of the inhibitors in common use inhibit photosynthesis and, therefore, the regeneration of the substrate for the photorespiratory pathway and are seldom specific or effective enough to permit unequivocal interpretations.[9]

The most recent attempt to measure flux through the photorespiratory pathway is a variation of the inhibitor method which circumvents some of the problems inherent in the use of inhibitors. A photorespiratory mutant of *Arabidopsis* lacking activity for the mitochondrial enzyme serine transhydroxymethylase cannot metabolize glycine to serine, and glycine accumulates as a stable endproduct of the photorespiratory pathway. Normally, photosynthesis is inhibited in this mutant in atmospheres stimulating photorespiration. However, photosynthesis rates can be restored by supplementing leaves with serine and NH_4Cl. Because these mutants do not evolve photorespiratory CO_2, rates of true photosynthesis in the absence of photorespiratory CO_2 release and refixation can be determined by measuring $^{12}CO_2$ uptake with an infrared gas analyzer. In this system, photorespiratory CO_2 evolution was about 27% of true photosynthesis (this is equivalent to about 1.4 μmol carbon entering the photorespiratory pathway per micromole apparent CO_2 fixed).[155]

All methods indicate that the flux of carbon into the photorespiratory pathway is approximately equal to the rate of carbon fixed in the Calvin cycle (true photosynthesis).[155] For comparison, the amount of NH_3 refixation associated with photorespiration exceeds primary NO_3^- reduction 10- to 50-fold,[167] and synthesis of photorespiratory glycine may exceed glycine utilization in protein synthesis about 1000- to 10000-fold.

VI. REGULATION OF PHOTORESPIRATION

A. Effects of Environmental Conditions

Numerous studies have established in a qualitative sense that glycolate synthesis is responsive to environmental regulation.[6-13] In general, low pCO_2, high pO_2, high temperature, and high light stimulate glycolate formation and flow through the photorespiratory pathway. The effects of pO_2 and pCO_2 are readily explained by their effects on the ratio of RuBP oxygenase to carboxylase activity as described in an earlier section. The effect of temperature appears to be due to the differential solubility response of O_2 and CO_2 to temperature,[67,68] and possibly by a differential effect of temperature on the kinetic properties of RubisCO.[6,17] The response to light intensity is presumably due to the fact that the absolute amount of P-glycolate production is proportional to the amount of RuBP available.[84] The possibility of RuBP limitation appeals to common sense but has been difficult to establish experimentally. Measurements of the effect of CO_2 concentration on RuBP concentration failed to reveal a significant effect, although the presence of O_2 did reduce the level of RuBP dramatically at very low CO_2 concentrations.[181] However, because the concentration of RuBP in the chloroplast may be less than one tenth the concentration of active sites on RubisCO,[182] the RuBP pool must turn over extremely quickly. It seems likely that the methods used to obtain measurements of this metabolite have been inadequate.

Recently, it has been noted that the effects of O_2 and CO_2 on glycolate synthesis are not entirely opposite in effect. The amount of label which accumulated in glycolate following chemical inhibition of the glycolate oxidase reaction in isolated cells was maximal at intermediate CO_2 concentrations.[100,183] The most likely explanation is that a certain amount of CO_2 fixation is necessary to maintain optimal levels of RubisCO activation or RuBP synthesis.

The effects of CO_2 and O_2 on photorespiratory CO_2 release (as distinguished from glycolate synthesis per se) have been less thoroughly examined because of the technical difficulties noted earlier. Analysis of the effect of CO_2 concentration on photorespiratory CO_2-release made by the dual carbon isotope method[52] have not been consistent with the other approaches. Little or no diminution of photorespiratory CO_2 release was observed as the CO_2 concentration of air was increased up to about 1200 μl l^{-1}.[65,66]

In an attempt to resolve this discrepancy, we employed a mutant of *Arabidopsis* in which true and apparent photosynthesis were identical because the mutant lacked photorespiratory CO_2 evolution. The effect of O_2 and CO_2 on photorespiration was then measured by quantitating the amount of glycine which accumulated at the blocked glycine decarboxylation step.[155] The results of these experiments demonstrated that the ratio of glycine accumulated to CO_2 fixed is a linear function of the O_2 concentration. The inhibitory effect of CO_2 on glycine synthesis was very pronounced (Figure 3), although relatively high (>0.3%) concentrations were necessary to prevent flow of carbon into the pathway. Thus, the apparent lack of effect of CO_2 concentration on photorespiration in the experiments employing the dual isotope method[52,65,66] are considered to be due to anomalous gas exchange phenomena rather than to an error in our perception of the biochemistry of photorespiration. It seems possible that a variable amount of internal CO_2-refixation in response to CO_2 concentration is responsible for the effect.[67,68]

An approximate ratio of oxygenation to carboxylation in vivo can be calculated from the various measures of photorespiration.[6,7,17] For instance, we have measured rates of photorespiratory flux in standard atmospheric conditions which indicate that one RuBP was oxygenated for each two molecules carboxylated. This has the net effect of reducing photosynthesis by 33% due to competition for RuBP. Also, since each oxygenation normally leads to the release of $^1/_2$ molecule of CO_2 as photorespiration, the remaining photosynthesis would have been reduced by an additional 25% because of the counteractive effects of photorespiration on photosynthesis. Thus, for each three carbons which would be fixed in the absence of O_2, only 1.5 are actually fixed (net). RuBP oxygenase activity, therefore, reduces photosynthesis by 50%.

Thus, there are, in principle, two different levels at which one may completely or partially overcome the effects of the properties of RubisCO on photosynthesis. One can, in principle, eliminate both O_2-inhibition of photosynthesis and photorespiration

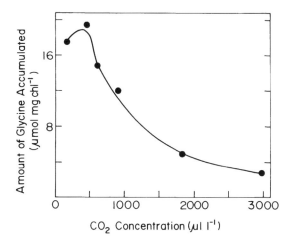

FIGURE 3. The effect of atmospheric CO_2 concentration on the amount of carbon entering the photorespiratory pathway. Leaf fragments of a mutant of *Arabidopsis* which lacked photorespiration and was unable to metabolize glycine because of a deficiency in mitochondrial serine transhydroxymethylase activity were illuminated in air containing 21% O_2 and the indicated amount of CO_2. After 15 min, the amount of glycine which accumulated was measured.[155]

by eliminating RuBP oxygenase. Alternately, one may envision altering the metabolism of P-glycolate so as to prevent the release of CO_2.

B. Effects of Metabolites or Inhibitors

The concept that photorespiration can be reduced by a modification of photorespiratory carbon metabolism[175,184] is an idea that preceded the discovery of the mechanism responsible for photorespiration. In this respect, it is based on a desirable end result rather than a direct insight into the nature of the problem. The early reports that inhibition of the enzymes of the photorespiratory pathway resulted in enhanced photosynthesis[184] provided a useful stimulus to research. However, the accumulated evidence strongly suggests that genetic or chemical modification of the pathway is actually deleterious.[185] A strong reduction in photosynthesis is observed when the photorespiratory pathway is blocked with α-HPMS,[98-101,186] INH,[100,101] HBA,[100-102] or AAN.[149,150]

Similar results have been obtained by genetic analysis. Somerville and Ogren have described mutants deficient in P-glycolate phosphatase,[82] serine:glyoxylate aminotransferase,[108] serine transhydroxymethylase,[47] glycine decarboxylase,[128] glutamate synthase,[171] and the chloroplast dicarboxylate transporter.[170] Each of these mutants grows normally in atmospheres enriched with concentrations of CO_2 which prevent RuBP oxygenase activity (i.e., 1% CO_2 in air). Also, when illuminated in atmospheres containing low O_2 (i.e., 350 μM CO_2 l^{-1}, 2% O_2), photosynthesis is normal and not distinguishable from that of the wild-type. However, when exposed to conditions which permit RuBP oxygenation, photosynthesis is dramatically inhibited. The reason why these mutations, and presumably chemical inhibitors, reduce photosynthesis varies. In the case of the P-glycolate phosphatase mutant, the accumulation of P-glycolate is believed to inhibit triose phosphate isomerase, thereby imposing a *de facto* block in the Calvin cycle.[82] In the case of the other mutants, there are two effects. First, the accumulation of glycine or serine, and the inability to recycle photorespiratory ammonia, leads to depletion of amino donors for glyoxylate transamination. As a result, glyoxylate is decarboxylated at rates which effectively counteract net photosynthesis. This loss of carbon which is normally recycled to the Calvin cycle may deplete the level of Calvin cycle intermediates.[108]

We have recently obtained evidence for the importance of flux through the terminal steps of the pathway from serine to PGA by establishing conditions which permit maintenance of normal photosynthesis rates in a mutant lacking serine transhydroxymethylase activity. By providing exogenous serine and NH_3, normal photosynthesis rates were obtained in atmospheric conditions which would otherwise have caused severe inhibition of photosynthesis.[155] The simple interpretation that photosynthesis was restored by ammonia-dependent synthesis of glutamate and serine-dependent reversal of Calvin cycle depletion may not be entirely correct. Créach and Stewart[150] have observed that leaf tissue treated with the glycine decarboxylase inhibitor AAN, has photosynthetic properties very similar to mutants of *Arabidopsis* which are deficient in glycine metabolism. In particular, photosynthesis is strongly inhibited in normal atmospheric conditions. However, under conditions where photosynthesis is impaired, RuBP levels are apparently normal. The implication is that RubisCO is in some way inhibited. Créach and Stewart[150] suggested that a slight accumulation of glyoxylate may inhibit RubisCO directly under these circumstances. In support of this it has been reported that glyoxylate is a very potent inhibitor of RubisCO.[187]

One of the most vigorously advanced "photorespiratory regulators" is glycidate (2,3-epoxypropionic acid), which has been advocated as an inhibitor of glycolate synthesis.[124,175] The claim is that application of this compound caused a 40 to 50% inhibition of glycolate synthesis which was accompanied by a 40 to 50% increase in photosynthetic CO_2 uptake. The proposed site of action is inhibition of glutamate: glyoxylate aminotransferase,[188] and NADPH-glyoxylate reductase.[124] However, on the basis of published data, we suggest that the compound is at best a "very weak" inhibitor of either activity.

Kumarasinghe et al.[101] examined the effect of glycidate in some detail and found that it actually inhibits photosynthesis. Chollet[64] also found that glycidate inhibited photosynthesis and had no differential effect on photorespiration. In examining the basis for the discrepancy, he obtained convincing evidence that the measure of photorespiration employed by Zelitch[63] was inappropriate, as suggested by others.[7,40,52] In view of the fact that the mode of glycidate action proposed by Zelitch[124,175] is not readily reconciled with the model of photorespiration presented here, and the inability of others[64,101] to observe any beneficial effect, we suggest that the significance of glycidate should be viewed skeptically. There have been two anomalous observations in which glycidate enhanced the rate of CO_2 fixation by isolated chloroplasts, but the effect was apparently not related to glycolate and/or photorespiratory metabolism.[189,190]

A common mechanism of metabolic regulation in microorganisms is end-product inhibition or allosteric regulation. In the simplest case, the end product of a metabolic pathway regulates its own synthesis by inhibiting the activity of the first enzyme of the pathway, thereby preventing excessive production of the end product. This concept has been of limited application in eukaryotic cells. In part, this reflects a dramatic difference in the organization of eukaryotic cells into many separate spaces bounded by semipermeable organellar membranes. The photorespiratory pathway, in particular, is not likely to be subject to metabolite regulation because of the partitioning of reactions in three organelles. Nonetheless, there are reports of such regulation. Oliver and Zelitch[191] reported that floating tobacco leaf discs on glyoxylate inhibited photorespiration and stimulated photosynthesis by a factor of two. An undefined feedback mechanism was proposed as the probable basis. However, glyoxylate was apparently a potent inhibitor of photosynthesis by isolated, intact chloroplasts — an unexplained discrepancy with the reported effect on intact tissue. In subsequent studies, it was reported that glyoxylate alleviated the O_2 inhibition of photosynthesis in leaf fragments[192] and with isolated soybean cells.[193] Oliver and Zelitch[194] also reported that floating leaf discs on glutamate inhibited photorespiration and stimulated photosynthesis. No specific mechanism to explain these observations was proposed.

Several other laboratories have examined the effects of glyoxylate. Chollet[64] reproduced the experimental observations reported by Oliver and Zelitch[191] but could see no inhibition of photorespiration or stimulation of photosynthesis by conventional means of measurement. Baumann and Gunther[195] did not observe any increase in photosynthesis following glyoxylate treatment of *Chenopodium album* cells, and Yun et al.[196] did not observe any beneficial effect in rice. Thus, the enthusiasm of some authors for metabolic regulation of photorespiration is not sustained by reproducibility. As noted previously, glyoxylate is extremely reactive and unstable. It inhibits three enzymes within the Krebs cycle by forming toxic conjugates with Krebs cycle intermediates,[197] inhibits photosynthesis of isolated chloroplasts,[192] inhibits the mitochondrial reactions associated with glycine decarboxylation,[198] and inhibits RubisCO.[187] Until the reports of a beneficial effect of glyoxylate on photosynthesis are widely reproduced and a rational mode of action proposed, it should be considered at most a potentially interesting anomaly.

We consider the concept of metabolic regulation of photorespiration to be wishful thinking. Once carbon enters the pathway, it moves forward by a series of irreversible reactions. The only substantially reversible steps are serine transhydroxymethylase and hydroxypyruvate reductase, and in both cases the equilibrium is maintained in the forward direction during photosynthesis by the availability of the cofactors C_1-THF and NADH, respectively. The importance of this is that mass action effects do not appear to play an important role in regulating the flow of carbon through the pathway. None of the photorespiratory intermediates have any regulatory effect on RuBP oxygenase activity and, for this reason, all claims for regulation by metabolites necessarily involve the otherwise undemonstrated existence of some other pathway of glycolate synthesis that is subject to regulation. We conclude that photorespiration cannot be regulated by modifying the pathway of glycolate synthesis. The only way of reducing photorespiration and the O_2 inhibition of photosynthesis is to modify RubisCO so that it does not produce P-glycolate.[185]

VII. RUBISCO

The relatively vast literature concerning RubisCO has recently been condensed into several reviews.[9,10,182,199] However, as the properties of the enzyme are central to any discussion concerning regulation of photorespiration, a brief discussion is warranted in the present context (see also Chapter 5).

A. Practical Considerations

There are several properties of RubisCO which make it attractive as an object of experimental interest, and several others which render analytical work with the enzyme technically intimidating. One of the attractive features is the abundance in leaves of C_3 species, which varies from 25 to 60% of total soluble leaf protein.[200] Because of the abundance and the high molecular weight of the native enzyme (*ca.* 550,000) it can be purified from a crude extract in several hours by centrifugation in vertical gradients of sucrose.[201]

On the other hand, two of the substrates are ubiquitous gases and one of the substrates (CO_2) is also a regulatory metabolite with a distinct binding site.[9] Thus, relatively elaborate precautions are necessary during the preparation of reagents and the assay to ensure control of substrate concentrations and enzyme activation.[202-204] Even the size of reaction containers may exert dramatic effects because of the partitioning of substrates between the gas and liquid phases.[58] The appropriate choice of pK^1 value for the CO_2 - HCO_3^- interconversion may alter calculated Km (CO_2) values dramatically,[204-206] and the differential effect of temperature on O_2 and CO_2 solubility ratios must also be considered.[67,68] Ideally, measurements of relative amounts of carboxylase and oxygenase activities must be conducted simultaneously. This is most readily accomplished by employing radioactive RuBP,[207,208] which is not commercially available. Although these and several other important considerations have not received uniform attention in the past, it is hoped that future studies involving the enzyme will be improved by more widespread attention to the various artifacts and errors which arise from inappropriate methodology.[182]

B. Molecular Biology

RubisCO is found exclusively in the chloroplasts of higher plants as an aggregate of about 550 kdaltons, composed of eight large subunits (LS) of approximately 55 kdaltons and eight small subunits (SS) of 12 to 18 kdaltons.[182,199] The subunits may be dissociated and separated only under relatively harsh denaturing conditions. Isolated small subunits lack catalytic activity but the large subunit has been reported to retain a low level of activity.[199] It has been generally accepted, therefore, that the LS contains the structural features necessary for catalysis. The SS is believed to enhance activity by altering the conformation of the LS toward more efficient catalysis,[205] rather than contributing to catalytic function.

The LS is encoded by the chloroplast genome[209-211] and the gene encoding the LS has been cloned and sequenced from several species.[211,212] The SS is encoded by the nuclear genome[210] and its mRNA is translated on cytoplasmic ribosomes.[213] The SS mRNA encodes a precursor polypeptide which contains an N-terminal sequence of amino acids not found in the mature SS. This transit sequence is believed to be involved in facilitating the transport of the SS across the chloroplast envelope.[213,214] At some stage in this process, the N-terminal sequence is cleaved off, presumably by a sequence-specific endopeptidase. Very little is known about the sequence of events leading to assembly of the mature S_8L_8 protein, but additional assembly proteins may be involved.[215] DNA sequences complementary to the mRNAs encoding the SS from pea have been cloned by several groups[216,217] and preliminary analyses suggest the presence of single or only a few copies of the gene per haploid genome.[217]

C. Enzymology

Since, as noted in an earlier section, RuBP oxygenation is believed to be the sole source of photorespiratory carbon, the question concerning modification of photorespiration and the O_2 inhibition of photosynthesis ultimately devolves to a question concerning the "linkage" of the carboxylase and oxygenase reactions. Can RuBP oxygenation activity be reduced without adversely affecting RuBP carboxylase activity?

The linearly competitive inhibition of carboxylation by O_2 and of oxygenation by CO_2 has been interpreted to suggest that CO_2 and O_2 compete with one another for a common enzyme-bound species of RuBP, or for a different species in rapid equilibrium with one another.[6,16] Thus, both activities involve the same active site and the same substrate. This is also suggested by the fact that CO_2 is an "activator" of both oxygenase and carboxylase activities.[202,218] Some difference in the mechanism underlying the two activities may be indicated by the observation that substitution of Mg^{2+} with Mn^{2+} or Co^{2+} changes the ratio of carboxylase to oxygenase activity.[208,219-221] However, the mechanistic implications of this observation is unknown.[9] A number of other reports which claimed differential reduction of RuBP oxygenase activity by various chemicals or enzymes added to the assay mixture[222-226] were not reproducible[208] and are considered artifacts of inappropriate methodology.

In the absence of an experimentally substantiated model for the chemical mechanisms responsible for the catalytic activities of RubisCO,[9] it is not possible to attribute significance to the differential effects of metals on the two activities. Such studies are of significance only inasmuch as they eventually lead to the development of such a model, or to a direct contradiction of what may be considered the "null hypothesis". Andrews and Lorimer[227] have proposed that RuBP oxygenase activity is an "inevitable consequence" of the chemistry of the RuBP carboxylation reaction mechanism. They proposed that at an intermediate stage of the carboxylation reaction, an enzyme-bound derivative of RuBP may be susceptible to attack by a species of oxygen. In this scheme, oxygenase activity is essentially a nonenzymatic side reaction since it is not necessary to postulate that oxygen interacts directly with the enzyme itself. One of the possible implications is that there is no "oxygen-binding site" which may be subject to modification by genetic or chemical means. This elegant concept provides a plausible explanation for the ubiquitous coincidence of RuBP carboxylase and oxygenase activity in all photosynthetic organisms examined to date. However, a useful hypothesis does not merely explain things away; it suggests experimental approaches. In this respect, the concept of inevitability is of limited utility. It does not, for instance, provide any insight into what properties of the active site topology or the chemical properties of the reactants may determine the relative rates of the two activities. It does not predict whether we should expect a constant ratio of the two activities in all organisms. If the ratio is not constant in all sources of the enzyme, then there is no compelling reason to believe that the ratio cannot be further changed in some beneficial way.

D. Variation in the RuBP Oxygenase/Carboxylase Ratio

Several recent studies may be interpreted as evidence that the kinetic constants of RubisCO may be amenable to directed genetic modification. Yeoh et al.[228,229] have compared $K_m(CO_2)$ and $K_m(RuBP)$ values among members of a large, taxonomically diverse collection of species. The $K_m(CO_2)$ values of C_3 species ranged from 12 to 25 μM, and the average value was less than half the average value found in C_4 species. The $K_m(RuBP)$ varied from 10 to 136 μM. In a similar study, Jordan and Ogren[230] employed a simultaneous assay of carboxylase and oxygenase activities to compare the ratio of activities in species representative of several major groups of photosynthetic organisms. Their studies revealed dramatic (up to ninefold) differences in the ratio of the two activities expressed as a function of both K_m and V_{max} values. The major cause of variation was change in $K_m(CO_2)$. However, substantial differences in $K_m(O_2)$ and V_{max} values were also observed. Bearing in mind the difficulties associated with comparing V_{max} values, these results strongly suggest that the ratio of the two activities is not immutable.

A similar conclusion was reached by Bird et al.[231] who compared $K_m(CO_2)$ values of RubisCO from a number of species. As in the studies of Yeoh et al.[228,229] and Jordan and Ogren,[230] significant differences were observed in $K_m(CO_2)$ among species. The similarity of the conclusions reached on three continents suggests that significant differences between species in the apparent Michaelis-Menton constants of RubisCO do exist and that assay procedures for RubisCO have developed to the point where such variation can be detected.

In contrast to the results of interspecies comparisons, attempts to identify variation within species have been unsuccessful and fraught with irreproducibility. A report[232] that a diploid and a derivative tetraploid form of ryegrass (*Lolium perenne* L.) had different affinities for CO_2 was reinvestigated in several laboratories but was not reproducible.[206,233,234] A report that the carboxylase/oxygenase ratio in tomato fruit changes during development[235] was also not substantiated.[36] The heterozygous tobacco *aurea* mutant *Su/su* has been reported to have higher rates of photorespiration than the wild-type *(su/su)*.[237,238] Although no difference in RubisCO was apparent by iso-electric focusing, it was reported that the enzyme from the mutant had an altered ratio of carboxylase to oxygenase activity.[239] The properties of the RubisCO from the mu-tant were reexamined in detail by Koivuniemi et al.[240] who could find no significant difference in the kinetic or physical properties of the enzyme from the mutant as com-pared to the wild-type. They noted that substantial variations in specific activity and physical properties of the enzyme are readily found as artifacts of enzyme preparation. Also, the reports of altered photorespiration in the *aurea* mutant of tobacco[238] were based on the use of a method for measuring photorespiration that has been strongly criticized.[64] Thus, we conclude that there is no substantiated report of natural variation of the ratio of carboxylase to oxygenase activity within a species. In view of the sub-stantial effort that has been wasted in reexamining the various reports of variability, it is to be hoped that more critical primary analyses will accompany any future reports of within-species variation.

Several lines of evidence suggest that significant natural diversity of the catalytic properties of RubisCO within a species may be extremely rare. A comparison of the isoelectric point of RubisCO from 39 species of *Triticum* and *Aegilops* revealed only two isoelectric variants of the large subunit.[241] Chen et al.[209] previously made a similar observation using a more limited collection of *Triticum* species and suggested that, because of the high degree of conservation of the physical properties of RubisCO, it may be a useful marker for phylogenetic studies. Similar studies with *Nicotiana*,[210] *Avena*,[242] *Petunia*,[243] and *Brassica*[244] have revealed similar high levels of apparent con-servation. A high degree of conservation between species is also indicated by immu-nochemical comparisons in which antigenic similarities were observed in all 50 species of angiosperms and gymnosperms studied.[245]

The most compelling evidence for the conservation of LS composition is the appar-ent conservation of amino acid sequences deduced from the DNA sequences of the maize and spinach enzymes,[211,212] and the partial amino acid sequences of barley.[246] Of 475 amino acids comprising the maize and spinach large subunits, only 44 are different. Of the 278 amino acids in the known barley sequences, only 7 are different from the maize sequence.[246] This may indicate that the multiplicity of chloroplast genomes con-tained in plant cells is a very conservative force in evolution.

Variation has been induced in RubisCO in microorganisms. A maternally inherited mutation in *Chlamydomonas rheinhardtii* eliminated both RuBP carboxylase and ox-ygenation activities.[47] This mutant is, of course, unable to photosynthesize and must be grown on an exogenous carbon source. Mutations which reduce RubisCO activity have also been isolated in the bacterium *Alcaligenes eutrophus.*[248] Several mutants were isolated, having a wide range of catalytic activities. In each case, the carboxylase and oxygenase activity declined to the same extent, substantiating biochemical evidence that the two activities are intricately and tightly linked.

The amino acid composition of the SS of pea and spinach have been determined[249,250] and N-terminal amino acid sequences are available for several other species.[251,252] These studies indicate that the amino acid composition of the SS or RubisCO is appar-ently less stringently conserved than the LS.[249] In view of the uncertainty concerning the function of this subunit,[205] it is possible that useful variability exists for this pro-tein. Rhodes et al.[253] have analyzed the enzymes produced by lines of *Nicotiana* in which genes encoding electrophoretically distinguishable SS are introduced into geno-types with electrophoretically distinguishable LS. Although no statistically significant differences were observed in the ratio of carboxylase to oxygenase activities, potential differences were obtained in the specific activity of both. Such studies are unlikely to eliminate photorespiration but may provide some insight into the function of the SS.

VIII. THE SEARCH FOR PLANTS WITH REDUCED PHOTORESPIRATION

A. Variation in C$_3$ Species

Because of the controversy concerning the precise mechanisms involved in photorespiration, there have been a number of attempts to sidestep the dangers associated with adherence to a flawed theory by searching directly for the desired characteristic. For instance, Lloyd and Canvin[176] performed detailed gas exchange measurements on 107 sunflower selections for variation in photosynthesis and photorespiration. As expected, on the basis of the properties of RubisCO, photosynthesis and photorespiration were positively correlated. No potentially useful variation was found. In an attempt to increase the possibility of encountering useful or informative variability, McCashin and Canvin[76] performed detailed gas exchange measurements on 26 mutant lines of barley which had previously been characterized as having altered photosynthetic characteristics. All measurements of photorespiration showed that photorespiration was positively and linearly correlated with photosynthesis. They concluded that the possibility of finding useful variation by direct screening was low.

These studies emphasize the point, which may also be concluded from studies of the physical properties of RubisCO, that in order to find variability, a screening system must be devised which permits evaluation of a large number of individuals. For this reason, the screening system should not depend on quantitative measurements of individual plants. An elegant and theoretically sound approach which satisfies this criterion was devised to exploit the fact that when C$_3$ plants are illuminated in atmospheres containing CO_2 concentrations below the CO_2 compensation point, net loss of CO_2 occurs due to photorespiration. If C$_4$ plants (compensation point less than 10 μl CO_2 1^{-1}) are enclosed with C$_3$ plants, the C$_4$ plants drive the CO_2 concentration below the compensation point of the C$_3$ plants. Under such circumstances, the C$_3$ plants begin to senesce after several days due to net loss of carbon, whereas the C$_4$ plants remain green and relatively healthy.[254,255] This approach has been used to screen many thousands of seedlings from heavily mutagenized lines of wheat and oats, as well as diverse genotypes of barley, soybeans, potatoes, and *Arabidopsis*.[254-257] No useful variation has been recovered from such screens, although some variability in length of survival was reported.[257] Recognizing that the all or none approach might be inappropriate, Nelson et al.[258] screened 6600 seedlings of *Festuca arundinacea*, Schreb. and saved the 58 most long-lived. No heritable differences were found in photosynthesis or photorespiration, but several plants appeared to have reduced rates of dark respiration. They suggested that variability in time of survival could be related primarily to seedling size and variability in reserve carbohydrate at the onset of the screen. We know of no inherent flaw in the CO_2 compensation point screening method. Since it has failed, we must assume that no significant variation was encountered.

By far the largest screening program for mutants with reduced photorespiration has been conducted using mutagenized populations of the small crucifer *Arabidopsis thaliana*.[185] Mutants were first characterized which lacked specific enzyme activities associated with photorespiratory metabolism. These mutants were able to grow normally in atmospheres enriched with CO_2 (which suppress RuBP oxygenation), but died after several days of exposure to air because of the toxic accumulation of photorespiratory intermediates and related effects.[155,185] The inability of these mutants to grow in air provided the basis for a screening method to detect secondary mutations which, by preventing flow of carbon into the photorespiratory pathway, would permit the mutant lines to survive in normal atmospheric conditions. Approximately five million heavily mutagenized plants were screened for survivors without recovering a plant with "reduced photorespiration". Several conclusions may be reached from these studies.

First, it does not appear to be possible to select for alternate metabolic fates for photorespiratory intermediates such as P-glycolate, serine, or glycine. Second, if it is possible to obtain a mutation which will prevent flow of carbon into the photorespiratory pathway (i.e., an RuBP oxygenase-deficient mutant), such mutants are exceedingly rare. The approach has been discontinued as unproductive.

The possibility of identifying useful variation by tissue culture methodology has also been explored.[259-262] Twenty stable variant lines, resistant to the inhibitor INH, were isolated in photoautotrophic cell cultures of *Nicotiana tabaccum* and resistant plants regenerated.[260] The rationale for the approach appears to have been based on the concept that by blocking the metabolism of glycine, one might force other, less wasteful, routes of glycine metabolism. In view of our inability to isolate a mutant of *Arabidopsis* which can metabolize glycine by some other route,[185] we consider this approach futile.

The INH resistant mutants have abnormalities in leaf morphology and plant size that have prevented comparisons of photosynthesis and photorespiration. Zelitch and Berlyn[261] have presented preliminary results which suggest that mitochondria from the INH resistant calli show a slight (threefold) reduction of glycine decarboxylase inhibition by INH. Lawyer et al.[262] used another inhibitor, glycine hydroxymate (GH), to select resistant photoautotrophic cell lines. Although some of the regenerated plants yielded calli which were resistant, others did not. The pattern of inheritance was too complex to suggest any known pattern of nuclear inheritance. Berlyn[259] has also suggested that O_2 resistant photoautotrophic callus might be of interest. However, the rate of photosynthesis in photoautotrophic cell lines was less than 2 μmol CO_2 fixed g fwt^{-1} hr^{-1} which is about 1% of that found in leaves. It is, therefore, unlikely that photorespiratory metabolism is of sufficient magnitude in these tissues to provide the basis for a selection scheme related to photorespiratory characteristics.

B. C_3-C_4 Hybridizations

Soon after the discovery of C_4 photosynthesis, it became apparent that the biochemical and morphological differences between C_3 and C_4 species do not represent a deep evolutionary split in the plant kingdom. The origin of C_4 photosynthesis is polyphyletic and both C_3 and C_4 species can be present within a single genus. *Atriplex* was one of the first genera in which this was demonstrated. The relatively small genetic distance between these C_3 and C_4 species made possible the creation of an artificial hybrid between *A. rosea* (C_4) and *A. triangularis* (C_3).[263] Subsequently, 22 additional interspecific hybridizations have been attempted.[264] The only $C_3 \times C_4$ crosses which yielded fertile F_1 hybrids were those between the diploids *A. rosea* (C_4) × *A. triangularis* (C_3), and *A. rosea* × *A. glabriusculla* (C_3).[265] Because of abnormal chromosome pairing, members of F_2 populations had varying chromosome numbers which prevented a precise and detailed genetic analysis. However, F_1 hybrids were uniform, and of intermediate habit, leaf shape, and dentation. The F_1 hybrids also had a distinct bundle sheath surrounded by a radiate mesophyll. The hybrid was distinguished from the C_4 parent by several features more typical of C_3 anatomy. Overall, the F_1 was considered to be of "intermediate" morphology.[263] All of the enzymes known to be essential for C_4 photosynthesis were present in the F_1 hybrid, and the hybrid has isozymes from both parents but the activities were lower than in the C_4 parent. The labeling pattern of early products of $^{14}CO_2$ fixation was also intermediate,[266] suggesting partial operation of C_4 photosynthetic pathway in the F_1 hybrid. However, the photosynthetic performance of the F_1 was lower than either of the parents. This has been interpreted to mean that the compartmentation of the two CO_2 fixation cycles was incomplete in some respect, or that some component was lacking.[265] Some abnormality was evident from the fact that RubisCO was present in all chloroplast containing cells[267] at levels comparable to the C_3 parent.[266]

The chromosome numbers of the F_2 population varied from slightly above diploid to more than pentaploid. As might be expected, the F_2 and subsequent generations segregated for all C_3 and C_4 characteristics. Anatomical and biochemical characteristics of C_4 photosynthesis appeared to segregate independently. For this reason, the probability of recovering functionally C_4 individuals was considered remote. None of the approximately 300 F_2, F_3 and F_4 individuals screened were capable of functional C_4 photosynthesis.[265] It seems clear from the foregoing studies that although the genetic divergence between C_3 and C_4 species need not be great, many genes are involved. At least some of these genes are nuclear and unlinked. A great deal more work will have to be done to determine how many structural and regulatory genes exist before one can realistically assess the possibility of a useful hybridization. In this respect, studying the effect of single gene mutations which alter C_4 anatomy or photosynthesis might prove informative. For the foreseeable future, the hybridization approach is not a feasible strategy to reducing photorespiration in C_3 species.

C. Intermediate Species

C_4 species are considered to have evolved from C_3 species on many independent occasions and at least 15 genera contain both C_3 and C_4 types.[265] Perhaps it is not surprising, therefore, that several possible intermediate species have been identified in genera, such as *Panicum,* which contain both C_3 and C_4 types.[268,269]

P. milioides, the most thoroughly studied of these, exhibits the superficial morphological characteristics of C_4 species such as radial arrangement of mesophyll cells around vascular bundles and enlarged parenchyma cells with centripetally located chloroplasts.[269-271] The arrangement differs from a "typical" C_4 grass in having a greater number of highly vacuolated and relatively thin walled mesophyll cells between the bundle sheaths.

The gas exchange characteristics of the C_3-C_4 intermediate species clearly distinguishes them from C_3 or C_4 species. The intermediates are similar to C_3 species with respect to apparent rates of photosynthesis and measures of diffuse resistance. However, striking differences between the intermediates and C_3 species are observed in measurements of apparent photorespiration, the CO_2 compensation point, and O_2 inhibition of photosynthesis.[37,268,272,273] It is with respect to these latter characteristics that the unusual *Panicum* species appears to be quantitatively intermediate between C_3 and C_4. The intermediate types also show a reduced sensitivity to growth inhibition by O_2 as compared to C_3 species, but are more inhibited than C_4 species.[37]

Several aspects of the gas exchange characteristics suggest that the underlying difference between the C_3 and C_3-C_4 intermediates is qualitative rather than quantitative. For instance, in C_3 species, the CO_2-compensation point (Γ) increases linearly with pO_2 whereas in the C_3-C_4 intermediates, the response is nonlinear at low pO_2.[272] Another unusual effect is that in C_3 species Γ is relatively constant over a wide range of irradiance, whereas in the C_3-C_4 intermediates, Γ decreases by as much as 75% as irradiance is increased.[273] Similarly, inhibition of apparent photosynthesis by O_2 and apparent photorespiration varies dramatically in the intermediate species in response to changes in irradiance. These effects are not readily explained, for instance, by a beneficial change in the kinetic constants of RubisCO,[272] and may suggest instead that net CO_2 fixation in these species entails two complementary processes with different light response curves.

The biochemical basis for the reduced photorespiratory characteristics is not understood and has been the subject of some controversy. Rathnam and Chollet[274] reported the compartmentation of PEP carboxylase and RubisCO in the mesophyll and bundle sheath cells, respectively, and suggested that a portion of photosynthetic CO_2 fixation may occur by a C_4 pathway. This proposal was subsequently retracted because of some difficulty in reproducing critical observations.[275] Also, Ku et al.[276] were unable to find evidence of compartmentation, and Hattersley et al.[267] provided convincing immunological evidence that RubisCO is present in all chloroenchymatous cells. In addition, the products of photosynthetic $^{14}CO_2$-fixation are considered C_3-like.[277,278] The δ ^{13}C values of about 25°/$_{oo}$ are characteristic of C_3 species,[279] and no decrease in O_2 sensitivity of carboxylation efficiency is apparent in the intermediates.[280,281] Keck and Ogren[272] suggested that RubisCO from *P. milioides* had greater affinity for CO_2 and that this could account for the reduced O_2-inhibition of photosynthesis. However, recent analysis indicates that the Km (CO_2) of RubisCO from *P. milioides* is indistinguishable from many typical C_3 species.[228] It has also been suggested that the elevated PEP carboxylase activity found in the intermediate species may be involved.[272,277] However, higher PEP carboxylase activity per se would not be expected to lead to significantly reduced O_2 inhibition of apparent photosynthesis. Compartmentation of carboxylation reactions, as in C_4 species, is probably required.

There is no simple explanation for the characteristics of the intermediate types. Recent gas exchange analysis suggests that the reduced response of apparent photosynthesis to O_2 is due to reduced photorespiration rather than reduced inhibition of RuBP carboxylation by O_2.[273,281] Thus, the intermediates have some system for preventing decarboxylation of photorespiratory glycine or for more efficiently refixing photorespiratory CO_2.[278] Whatever the difference, it does not appear to be a CO_2 concentrating mechanism since it does not change the carboxylation efficiency.[273]

The *Laxa* group in *Panicum*, to which *P. milioides* belongs, contains about 15 to 20 tropical and subtropical species with C_3 intermediate types and possibly one C_4 species. Ploidy varies from diploid to hexaploid within the group, and may vary within species.[282] Preliminary attempts have been made to obtain hybrids between the C_3 and C_3-C_4 intermediates.[282] Although a few hybrids have been obtained, sterility has so far prevented analysis of subsequent generations. The F_1 has Γ values and rates of photorespiratory CO_2 loss intermediate between the two species.

IX. FUNCTIONAL SIGNIFICANCE OF PHOTORESPIRATION

A. Quantum Yield of Photosynthesis

Since C_4-species do not photorespire and generally have high rates of photosynthesis, one may legitimately ask why they have not predominated. A convincing answer to this question could have some relevance to attempts to reduce photorespiration. Are we trying to accomplish something which might ultimately prove deleterious? In the cooler and more temperate regions of North America, C_4 plants occur relatively infrequently. In contrast, C_4 plants are more prevalent in the deserts, grasslands, and other subtropical regions.[283] A possible basis for this effect is related to the quantum yield efficiency of C_3 and C_4 photosynthesis.[265,284] Under nonphotorespiratory conditions, C_3 plants have a greater quantum yield than C_4 plants because the C_4 adaptation requires more ATP per CO_2 fixed. However, as the amount of photorespiration increases in a C_3 species, the quantum efficiency decreases because of the greater costs associated with refixing photorespiratory CO_2 and NH_3, and regenerating RuBP. Under natural circumstances, the amount of photorespiration in a given C_3 plant increases with increasing temperature, in part, because of the differential effect of temperature on the solubility of CO_2 and O_2.[67,68] Lower temperature is associated with a lower dissolved O_2/CO_2 ratio and therefore lower photorespiration. Thus, when quantum yield of a C_3 species is plotted as a function of temperature, it decreases as temperature increases whereas the quantum yield of a C_4 species remains constant.[284] As a result, the quantum yield of C_3 species is higher than that of C_4 species below some critical temperature but lower above this temperature. In a comparison of several *Atriplex* species, the critical temperature was between about 25 and 30°C.[284] Thus, under conditions of adequate soil moisture and limiting light (as found in many canopies), a C_3 plant will have a greater potential for net carbon gain at lower temperature, whereas a C_4 would have the advantage at high temperature. In the absence of conflicting theories, this forceful hypothesis may partially explain the predominance of C_3 species in temperate niches. If this is a relevant argument, it can be concluded that it is not photorespiration which has been selected for, but the cost of suppressing photorespiration that has been selected against.

B. Photoinhibition

It is extremely difficult to posit a single selective force that might have maintained RuBP oxygenase activity in all photosynthetic organisms from photosynthetic anaerobes to higher plants. It seems likely, therefore, that RuBP oxygenase activity is ubiquitous for some reason related to the reaction mechanism of RuBP carboxylase rather than any desirable characteristic of the oxygenase reaction or the products thereof.[227] This does not mean, however, that RuBP oxygenase or dependent reactions serve no useful function. Gould and Lewontin[285] have eloquently argued that biological systems evolve useful employment for available parts, irrespective of the circumstances or forces which may have given rise to particular traits.

In contemplating possible functions for photorespiration, it is important to note that photorespiration is apparently dispensable to the well cared for C_3 plant. *Arabidopsis* can be grown, without any deleterious effect, throughout the life cycle in atmospheric conditions which prevent photorespiratory flux.[185] Thus, if photorespiration serves a function, it is an occasional one related to the variation in natural environments. An innovative scheme in this vein was proposed by Osmond and Björkman.[286] They suggested that under CO_2-limited conditions, such as drought-induced stomatal closure, photorespiration could provide a mechanism for orderly dissipation of photosynthetic reductant. The proposal is based on the idea that the only significant sink for photosynthetic reductant is the Calvin cycle. The consumption of ATP and NADPH by the Calvin cycle may, in principle, proceed by carboxylation or oxygenation of RuBP. Thus, in the absence of CO_2, RuBP oxygenase activity could permit Calvin cycle activity by consuming stored carbohydrate (which may be converted to RuBP) and by releasing and refixing photorespiratory CO_2.[11]

The protective function of photosynthetic carbon metabolism has been demonstrated by illuminating leaves, chloroplasts, or cells in atmospheric conditions which restrict RuBP oxygenase and carboxylase activity, either singly or simultaneously.[287-292] When leaves are illuminated in N_2 for several hours, the subsequent capacity for photosynthesis declines sharply due in part to a reduction in photosynthetic electron transport.[291,292] The presence of low concentrations of either CO_2 or O_2 partially or completely prevents this photoinhibition presumably by providing a sink for photoreductant via RuBP carboxylase and RuBP oxygenase activity, respectively.

Although the precise details of the mechanism of photoinhibition are uncertain, it seems clear that a certain amount of Calvin cycle activity is necessary during illumination. The remaining question is whether stomata of C_3 leaves are ever closed sufficiently to prevent totally the entry of CO_2. So far the evidence in this respect is inconclusive even when C_3 and C_4 plants are subject to the most extreme water stress.[11,265]

X. CONCLUSIONS AND RECOMMENDATIONS

The physiology and biochemistry of photorespiration has arrived at a condition of relative maturity. No new reactions have been discovered in more than a decade and the evidence supporting the theory of photorespiration based on the properties of

RubisCO and the subsequent metabolism of glycolate through the photorespiratory pathway is compelling. In spite of periodic claims to the contrary, there seems very little possibility that photorespiration can be modified in any useful way by altering the reactions of the pathway. The characteristics of mutants with defects in the pathway, and recent studies with chemical inhibitors, point to the fact that once P-glycolate is formed, it must be metabolized through to PGA in order to prevent disruption of photosynthetic CO_2 assimilation. There are two reactions within the pathway which must operate at maximal velocity in order to prevent excessive loss of carbon. These are glyoxylate transamination and hydroxypyruvate reduction. Nutritional or enzymatic deficiencies which limit these reactions will reduce net photosynthesis. However, there is no reason to believe that these are normally limiting. Therefore, attention must be focused on preventing P-glycolate formation.

The natural adaptations which suppress RuBP oxygenase activity involve indirect means such as the CO_2-concentrating mechanism of C_4 species. In view of the apparent genetic complexity of the C_4 syndrome, it is a conceptually daunting prospect to consider converting a C_3 to a C_4 type. Even assuming that techniques for gene transfer between species become of common-place utility, this approach does not seem feasible in the foreseeable future.

The uncertainty which surrounds the biochemical mechanisms responsible for the unusual photorespiratory characteristics of the C_3-C_4 intermediates prohibits a critical assessment of the potential significance of this phenomenon. Conceivably, there may be fewer genetic differences between the intermediates and C_3 species than between C_3 and C_4 species. However, since the intermediates are relatively rare, it seems possible that there is some disadvantage associated with the intermediate syndrome which has prevented it from becoming more widespread. More work needs to be done to assess the agronomic significance and the biochemical basis of this syndrome.

Although there have been many reports of variation within species for rate of photorespiration or the kinetic constants of RubisCO, there is very little reason for optimism. Many factors (i.e., storage carbohydrate, nutrition, or leaf architecture) may alter the rate of apparent photorespiration as measured by available methods without representing a heritable improvement in photosynthetic performance. Similarly, the assay of RubisCO is fraught with opportunities for error. Thus, in spite of the substantial labor involved world-wide in the search for useful variation, not a single variant plant has survived scrutiny in more than one laboratory. Because of the apparent scarcity of useful variability for this character, it seems absolutely futile to conduct gas exchange surveys of potential breeding material with an eye to reducing photorespiration.

Several theoretically sound and relatively large screening attempts which employed visual detection of desirable material have failed to identify useful material. Similarly, a protracted search for useful variability in tissue culture has not been profitable. The failure of these approaches substantiates the conclusion drawn from biochemical and physiological studies — there are no easy solutions to photorespiration.

This gloomy picture of failed and futile attempts does not mean that photorespiration is intractable. Rather, it suggests that genuinely new approaches are required. We believe that attention must be directed toward understanding the reaction mechanisms of RubisCO so that a rational evaluation of the possibility of altering the enzyme can be made on the basis of mechanistic rather than teleological criteria. If, as suggested by Andrews and Lorimer,[227] the oxygenase activity is inevitably associated with the chemistry of RuBP carboxylation, then attention must be directed toward identifying or creating enzymes with reduced Km (CO_2). Because the rate of oxygenase activity and the inhibitory effects of O_2 on carboxylation are inversely related to Km (CO_2), a decrease in this constant is an improvement of the enzyme. That such changes have already occurred now seems abundantly clear.[228-231] It now remains to ascertain what the limit of such changes may be. In addition to the conventional tools of the enzymologist, the techniques of molecular biology will be increasingly brought to bear on this aspect of the problem. At least six RubisCO LS genes have been cloned and are or will be sequenced in the near future. This will facilitate in vitro modification of the gene and correlations of differences in primary structure with differences in enzyme kinetics. Thus, there are exciting possibilities. In the short term, however, it would not seem worthwhile to include consideration of photorespiratory characteristics in a breeding program.

LIST OF ABBREVIATIONS

AAN; aminoacetonitrile
CH_2H_4-folate; N^5,N^{10}-methylene tetrahydrofolic acid
H_4-folate; tetrahydrofolic acid
GH; glycine hydroxymate
HBA; hydroxybutynoic acid
α-HPMS; α-hydroxy-pyridine methane sulfonic acid
INH; isonicotinic hydrazide
PAR; photosynthetically active radiation (i.e., radiation between 400 and 700 nm)
PEP; phosphoenol pyruvate
PGA; 3-phosphoglyceric acid
P-glycolate; 2-phosphoglycolic acid
RubisCO; ribulose-1,5-bisphosphate carboxylase/oxygenase (EC 4.1.1.39)
RuBP; ribulose-1,5-bisphosphate

REFERENCES

1. Decker, J. P., A rapid post-illumination deceleration of respiration in green leaves, *Plant Physiol.,* 30, 82, 1955.
2. Zelitch, I., Organic acids and respiration in photosynthetic tissues, *Annu. Rev. Plant Physiol.,* 15, 121, 1964.
3. Tolbert, N. E., Microbodies — peroxisomes and glyoxysomes, *Annu. Rev. Plant Physiol.,* 22, 45, 1971.
4. Ogren, W. L. and Bowes, G., Ribulose diphosphate carboxylase regulates soybean photorespiration, *Nature (London), New Biol.,* 230, 159, 1971.
5. Bowes, G., Ogren, W. L., and Hageman, R. H., Phosphoglycolate production catalyzed by ribulose diphosphate carboxylase, *Biochem. Biophys. Res. Commun.,* 45, 716, 1971.
6. Laing, W. A., Ogren, W. L., and Hageman, R. H., Regulation of soybean net photosynthetic CO_2 fixation by the interaction of CO_2, O_2, and ribulose 1,5-diphosphate carboxylase, *Plant Physiol.,* 54, 678, 1974.
7. Chollet, R. and Ogren, W. L., Regulation of photorespiration in C_3 and C_4 species, *Bot. Rev.,* 41, 137, 1975.
8. Schnarrenberger, C. and Fock, H., Interactions among organelles involved in photorespiration, in *Encyclopedia of Plant Physiology, New Series,* Vol. 3, Stocking, C. R. and Heber, U., Eds., Springer-Verlag, New York, 1976, 185.
9. Lorimer, G. H., The carboxylation and oxygenation of ribulose 1,5-bisphosphate: the primary events in photosynthesis and photorespiration, *Annu. Rev. Plant Physiol.,* 32, 349, 1981.
10. Lorimer, G. H. and Andrews, T. J., The C_2 chemo- and photorespiratory carbon oxidation cycle, in *The Biochemistry of Plants,* Vol. 8, Hatch, M. D. and Boardman, N. K., Eds., Academic Press, New York, 1981, 329.

11. Osmond, C. B., Photorespiration and photoinhibition. Some implications for the energetics of photosynthesis, *Biochim. Biophys. Acta*, 639, 77, 1981.
12. Tolbert, N. E., Photorespiration, in *The Biochemistry of Plants,* Vol. 2, Davies, D. D., Ed., Academic Press, New York, 1980, 487.
13. Canvin, D. T., Photorespiration: comparison between C_3 and C_4 plants, in *Encyclopedia of Plant Physiology, New Series,* Vol. 6, Gibbs, M. and Latzko, E., Eds., Springer-Verlag, New York, 1979, 368.
14. Andrews, T. J., Lorimer, G. H., and Tolbert, N. E., Ribulose diphosphate oxygenase. I. Synthesis of phosphoglycolate by fraction I protein of leaves, *Biochemistry*, 12, 11, 1973.
15. Lorimer, G. H., Andrews, T. J., and Tolbert, N. E., Ribulose diphosphate oxygenase. II. Further proof of reaction products and mechanisms of action, *Biochemistry*, 12, 18, 1973.
16. Badger, M. R. and Andrews, T. J., Effects of CO_2, O_2 and temperature on a high affinity form of ribulose diphosphate carboxylase-oxygenase from spinach, *Biochem. Biophys. Res. Commun.*, 60, 204, 1974.
17. Farquhar, G. D., Models describing the kinetics of ribulose bisphosphate carboxylase/oxygenase, *Arch. Biochem. Biophys.*, 193, 456, 1979.
18. Farquhar, G. D., von Caemmerer, S., and Berry, J. A., A biochemical model of photosynthesis, *Planta*, 149, 78, 1980.
19. Cooper, R. L. and Brun, W. A., Response of soybeans to a carbon dioxide-enriched atmosphere, *Crop Sci.*, 7, 455, 1967.
20. Hardman, L. L. and Brun, W. A., Effect of atmospheric carbon dioxide enrichment at different developmental stages on growth and yield components of soybeans, *Crop Sci.*, 11, 886, 1971.
21. Krenzer, E. G., Jr. and Moss, D. N., Carbon dioxide enrichment effects upon yield and yield components in wheat, *Crop Sci.*, 15, 71, 1975.
22. Fisher, R. A. and Aguilar, M., Yield potential in a dwarf spring wheat and the effect of carbon dioxide fertilization, *Agron. J.*, 68, 749, 1976.
23. Gifford, R. M., Growth pattern, carbon dioxide exchange and dry weight distribution in wheat growing under different photosynthetic environments, *Aust. J. Plant Physiol.*, 4, 99, 1977.
24. Gifford, R. M., Growth and yield of CO_2-enriched wheat under water limited conditions, *Aust. J. Plant Physiol.*, 6, 367, 1979.
25. Gifford, R. M., Brenner, P. M., and Jones, D. B., Assessing photosynthetic limitation to grain yield in a field crop, *Aust. J. Agric. Res.*, 24, 297, 1973.
26. Hardy, R. W. F. and Havelka, U. D., Possible routes to increase the conversion of solar energy to food and feed by grain legumes and cereal grains: CO_2 and N_2 fixation, foliar fertilization, and assimilate partitioning, in *Biological Solar Energy Conversion,* Mitsui, A., Miyachi, S., San Pietro, A., and Tamura, S., Eds., Academic Press, New York, 1977, 299.
27. Hardy, R. W. F. and Havelka, U. D., Nitrogen fixation research: a key to world food?, *Science*, 188, 633, 1975.
28. Finn, G. A. and Brun, W. A., Effect of atmospheric CO_2-enrichment on growth, non-structural carbohydrate content, and root nodule activity in soybean, *Plant Physiol.*, 69, 327, 1982.
29. Kramer, P. J., Carbon dioxide concentration, photosynthesis, and dry matter production, *BioScience*, 31, 29, 1981.
30. Zelitch, I., The close relationship between net photosynthesis and crop yield, *BioScience*, 32, 796, 1982.
31. Clough, J. M., Peet, M. M., and Kramer, P. J., Effects of high atmospheric CO_2 and sink size on rates of photosynthesis of a soybean culture, *Plant Physiol.*, 67, 1007, 1981.
32. Hicklenton, P. R. and Jolliffe, P. A., Alterations in the physiology of CO_2 exchange in tomato plants grown in CO_2-enriched atmospheres, *Can. J. Bot.*, 58, 2181, 1980.
33. Neales, T. F. and Nicholls, A. O., Growth response of young wheat plants to a range of ambient CO_2 levels, *Aust. J. Plant Physiol.*, 5, 45, 1978.
34. Björkman, O., Hiesey, W. M., Nobs, M., Nicholson, F., and Hart, R. W., Effect of oxygen concentration on dry matter production in higher plants, *Carnegie Inst. Washington Yearb.*, 66, 228, 1968.
35. Björkman, O., Gauhl, D. E., Hiesey, W. M., Nicholson, F., and Nobs, M. A., Growth of *Mimulus, Marchantia* and *Zea* under different oxygen and carbon dioxide levels, *Carnegie Inst. Washington Yearb.*, 67, 477, 1969.
36. Parkinson, K. J., Penman, H. L., and Tregunna, E. B., Growth of plants in different oxygen concentrations, *J. Exp. Bot.*, 25, 132, 1974.
37. Quebedeaux, B. and Chollet, R., Comparative growth analysis of *Panicum* species with differing rates of photorespiration, *Plant Physiol.*, 59, 42, 1977.
38. Quebedeaux, B. and Hardy, R. W. F., Oxygen concentration: regulation of crop growth and productivity, in *CO_2 Metabolism and Plant Productivity,* Burris, R. H. and Black, C. C., Eds., University Park Press, Baltimore, 1976, 185.

39. Sestak, Z., Catsky, J., and Jarvis, P. G., *Plant Photosynthetic Production: A Manual of Methods,* W. Junk, The Hague, 1971.
40. Canvin, D. T. and Fock, H., Measurement of photorespiration, *Methods Enzymol.,* 24, 246, 1972.
41. Jackson, W. A. and Volk, R. J., Photorespiration, *Annu. Rev. Plant Physiol.,* 21, 385, 1970.
42. Doehlert, D. C., Ku, M. S. B., and Edwards, G. E., Dependence of the post-illumination burst of CO_2 on temperature, light, CO_2, and O_2 concentration in wheat *(Triticum aestivum), Physiol. Plant,* 46, 299, 1979.
43. Bully, N. R. and Tregunna, E. B., Photorespiration and the post-illumination burst, *Can. J. Bot.,* 49, 1277, 1971.
44. D'Aoust, A. H. and Canvin, D. T., Caractéristiques du $^{14}CO_2$ dégagé à la lumière et a l'obscurité par les feuilles de haricot, de radis, de tabac et de tournesol pendant et après une photosynthese en présence de $^{14}CO_2$, *Physiol. Veg.,* 12, 545, 1974.
45. Forrester, M. L., Krotkov, G., and Nelson, C. D., Effect of oxygen on photosynthesis, and photorespiration in detached leaves. I. Soybean, *Plant Physiol.,* 41, 422, 1966.
46. Kisaki, T., Yoshida, N., and Imai, A., Glycine decarboxylase and serine formation in spinach leaf mitochondrial preparations with reference to photorespiration, *Plant Cell. Physiol.,* 12, 275, 1971.
47. Somerville, C. R. and Ogren, W. L., Photorespiration-deficient mutants of *Arabidopsis thaliana* lacking serine transhydroxymethylase activity, *Plant Physiol.,* 67, 666, 1981.
48. Vines, H. M., Armitage, A. M., Chen, S.-S., Tu, Z.-P., and Black, C. C., A transient burst of CO_2 from geranium leaves during illumination at various light intensities as a measure of photorespiration, *Plant Physiol.,* 70, 629, 1982.
49. Hew, C. S. and Krotkov, G., Effect of oxygen on the rates of CO_2 evolution in light and in darkness by photosynthesizing and non-photosynthesizing leaves, *Plant Physiol.,* 43, 464, 1968.
50. Cornic, G., Étude du dégagement de CO_2 a la lumière dans un air dépourvu de gaz carbonique chez la moutarde blanche *(Sinapsis alba* L.), *Physiol. Veg.,* 10, 605, 1972.
51. Cornic, G., Etude de l'effet de la temperature sur la photorespiration de la moutarde blanche *(Sinapsis alba* L.), *Physiol. Vég.,* 12, 83, 1974.
52. Ludwig, L. J. and Canvin, D. T., The rate of photorespiration during photosynthesis and the relationship of the substrate of light respiration to the products of photosynthesis in sunflower leaves, *Plant Physiol.,* 48, 712, 1971.
53. D'Aoust, A. L. and Canvin, D. T., The specific activity of $^{14}CO_2$ evolved in CO_2-free air in the light and darkness by sunflower leaves following periods of photosynthesis in $^{14}CO_2$, *Photosynthetica,* 6, 150, 1972.
54. Brown, R. H., Photosynthesis of grass species differing in carbon dioxide fixation pathways. IV. Analysis of reduced oxygen response in *Panicum milioides* and *Panicum schenckii, Plant Physiol.,* 65, 346, 1980.
55. Hew, C., Krotkov, G., and Canvin, D. T., Determination of the rate of CO_2 evolution by green leaves in light, *Plant Physiol.,* 44, 662, 1969.
56. Smith, E. W., Tolbert, N. E., and Ku, H. S., Variables affecting the CO_2 compensation point, *Plant Physiol.,* 58, 143, 1976.
57. Krenzer, E. G., Jr., Moss, D. N., and Crookston, R. K., Carbon dioxide compensation points of flowering plants, *Plant Physiol.,* 56, 194, 1975.
58. Ogren, W. L. and Hunt, L. D., Comparative biochemistry of ribulose bisphosphate carboxylase in higher plants, in *Photosynthetic Carbon Assimilation,* Siegelman, H. W. and Hind, G., Eds., Plenum, New York, 1978, 127.
59. Bloom, A. J., Mooney, H. A., Björkman, O., and Berry, J., Materials and methods for carbon dioxide and water exchange analysis, *Plant Cell Environ.,* 3, 371, 1980.
60. Jolliffe, P. A. and Tregunna, E. B., Effects of temperature, CO_2 concentration, and light intensity on oxygen inhibition of photosynthesis in wheat leaves, *Plant Physiol.,* 43, 902, 1968.
61. Jolliffe, P. A. and Tregunna, E. B., Environmental regulation of the oxygen effect on apparent photosynthesis in wheat, *Can. J. Bot.,* 51, 841, 1973.
62. Canvin, D. T., Berry, J. A., Badger, M. R., Fock, H., and Osmond, B., Oxygen exchange in leaves in the light, *Plant Physiol.,* 66, 302, 1980.
63. Zelitch, I., Investigations on photorespiration with a sensitive ^{14}C-assay, *Plant Physiol.,* 43, 1829, 1968.
64. Chollet, R., Evaluation of the light/dark ^{14}C assay of photorespiration, Tobacco leaf disc studies with glycidate and glyoxylate, *Plant Physiol.,* 61, 929, 1978.
65. Bravdo, B. A. and Canvin, D. T., Effect of carbon dioxide on photorespiration, *Plant Physiol.,* 63, 399, 1979.
66. Fock, H., Klug, K., and Canvin, D. T., Effect of carbon dioxide and temperature on photosynthetic CO_2 uptake and photorespiratory CO_2 evolution in sunflower leaves, *Planta,* 145, 219, 1979.

67. Ku, S. and Edwards, G. E., Oxygen inhibition of photosynthesis. I. Temperature dependence and relation to O_2/CO_2 solubility ratio, *Plant Physiol.*, 59, 986, 1977.
68. Ku, S. and Edwards, G. E., Oxygen inhibition of photosynthesis, *Plant Physiol.*, 59, 991, 1977.
69. Brown, A. H., The effects of light on respiration using isotopically enriched oxygen, *Am. J. Bot.*, 40, 719, 1953.
70. Gerbaud, A. and Andre, M., Effect of CO_2, O_2, and light on photosynthesis and photorespiration in wheat, *Plant Physiol.*, 66, 1032, 1980.
71. Berry, J. A., Osmond, C. B., and Lorimer, G. H., Fixation of $^{18}O_2$ during photorespiration: kinetic and steady-state studies of the photorespiratory carbon oxidative cycle with intact leaves and isolated chloroplasts of C_3 plants, *Plant Physiol.*, 62, 954, 1978.
72. Radmer, R. J. and Kok, B., Photoreduction of O_2 primes and replaces CO_2 assimilation, *Plant Physiol.*, 58, 336, 1976.
73. Marsho, T. V., Behrens, P. W., and Radmer, R. J., Photosynthetic oxygen reduction in isolated intact chloroplasts and cells from spinach, *Plant Physiol.*, 64, 656, 1979.
74. Good, N. E. and Brown, A. H., The contribution of endogenous oxygen to the respiration of photosynthesizing *Chlorella* cells, *Biochim. Biophys. Acta*, 50, 544, 1961.
75. Farquhar, G. D. and Sharkey, T. D., Stomatal conductance and photosynthesis, *Annu. Rev. Plant Physiol.*, 33, 317, 1982.
76. McCashin, B. G. and Canvin, D. T., Photosynthetic and photorespiratory characteristics of mutants of *Hordeum vulgare* L., *Plant Physiol.*, 64, 354, 1979.
77. Jones, R. B. and Waley, S. G., Spectrophotometric studies on the interaction between triose phosphate isomerase and inhibitors, *Biochem. J.*, 179, 623, 1979.
78. Wolfenden, R., Binding of substrate and transition state analogs to triose phosphate isomerase, *Biochemistry*, 9, 3404, 1970.
79. Richardson, K. E. and Tolbert, N. E., Phosphoglycolic acid phosphatase, *J. Biol. Chem.*, 236, 1285, 1961.
80. Randall, D. D., Tolbert, N. E., and Gremel, D., 3-Phosphoglycerate phosphatase in plants. II. Distribution, physiological considerations and comparison with p-glycolate phosphatase, *Plant Physiol.*, 48, 480, 1971.
81. Christeller, J. T. and Tolbert, N. E., Phosphoglycolate phosphatase, *J. Biol. Chem.*, 253, 1780, 1978.
82. Somerville, C. R. and Ogren, W. L., A phosphoglycolate phosphatase deficient mutant of *Arabidopsis*, *Nature (London)*, 280, 833, 1979.
83. Verin-Vergeau, C., Baldy, P., and Cavalié, G., Les phosphoglycolate phosphatase des feuilles de haricot: propriétés catalytiques des deux isoenzymes, *Phytochemistry*, 19, 763, 1980.
84. Kirk, M. R. and Heber, U., Rates of synthesis and source of glycolate in intact chloroplasts, *Planta*, 132, 131, 1976.
85. Krause, G. H., Thorne, S. W., and Lorimer, G. H., Glycolate synthesis by intact chloroplasts, *Arch. Biochem. Biophys.*, 183, 471, 1977.
86. Heber, U. and Heldt, H. W., The chloroplast envelope: structure, function, and role in leaf metabolism, *Annu. Rev. Plant Physiol.*, 32, 139, 1981.
87. Takabe, T. and Akazawa, T., Mechanism of glycolate transport in spinach leaf chloroplasts, *Plant Physiol.*, 68, 1093, 1981.
88. Howitz, K. T. and McCarty, R. E., pH dependence and kinetics of glycolate uptake by intact pea chloroplasts, *Plant Physiol.*, 70, 949, 1982.
89. Liang, Z., Yu, C., and Huang, A. C., Isolation of spinach leaf peroxisomes in 0.25 Molar sucrose solution by Percoll density gradient centrifugation, *Plant Physiol.*, 70, 1210, 1982.
90. Schmitt, M. R. and Edwards, G. E., Isolation and purification of intact peroxisomes from green leaf tissue, *Plant Physiol.*, 70, 1213, 1982.
91. Kerr, M. W. and Groves, D., Purification and properties of glycolate oxidase from *Pisum sativum* leaves, *Phytochemistry*, 14, 359, 1975.
92. Fendrich, G. and Ghisla, S., Studies on glycollate oxidase from pea leaves: determination of sterospecificity and mode of inhibition by hydroxy-butynoate, *Biochim. Biophys. Acta*, 702, 242, 1982.
93. Frederick, S. E., Gruber, P. J., and Tolbert, N. E., The occurrence of glycolate dehydrogenase and glycolate oxidase in green plants, *Plant Physiol.*, 52, 318, 1973.
94. Beezley, B. B., Gruber, P. J., and Fredrick, S. E., Cytochemical localization of glycolate dehydrogenase in mitochondria of *Chlamydomonas*, *Plant Physiol.*, 58, 315, 1976.
95. Roth-Bejarano, N. and Lips, S. H., Glycolate dehydrogenase activity in higher plants, *Israel J. Bot.*, 22, 1, 1973.
96. Corbett, J. R. and Wright, B. J., Inhibition of glycolate oxidase as a rational way of designing a herbicide, *Phytochemistry*, 10, 2015, 1971.
97. Osmond, C. B. and Avadhani, P. N., Inhibition of the β-carboxylation pathway of CO_2 fixation by bisulfite compounds, *Plant Physiol.*, 45, 228, 1970.

98. Moss, D. N., Photorespiration and glycolate metabolism in tobacco leaves, *Crop Sci.*, 8, 71, 1968.
99. Latche, J. C., Bailly-Fenech, G., and Cavalie, G., Effets de l'α-hydroxy-2-pyridine methane sulfonate sur l'oxydation du glycolate et l'assimilation photosynthétique du gaz carbonique dans les feuilles de Soja, *C. R. Acad. Sci. Paris*, 293, 765, 1981.
100. Servaites, J. C. and Ogren, W. L., Chemical inhibition of the glycolate pathway in soybean leaf cells, *Plant Physiol.*, 60, 461, 1977.
101. Kumarasinghe, K. S., Keys, A. J., and Whittingham, C. P., Effects of certain inhibitors on photorespiration by wheat leaf segments, *J. Exp. Bot.*, 28, 1163, 1977.
102. Doravari, S. and Canvin, D. T., Effect of butyl 2-hydroxy-3-butynoate on sunflower leaf photosynthesis and photorespiration, *Plant Physiol.*, 66, 628, 1980.
103. Rehfeld, D. W. and Tolbert, N. E., Aminotransferases in peroxisomes from spinach leaves, *J. Biol. Chem.*, 247, 4803, 1972.
104. Noguchi, T. and Hayashi, S., Plant leaf alanine: 2-oxoglutarate amino transferase. Peroxisomal localization and identity with glutamate: glyoxylate aminotransferase, *Biochem. J.*, 195, 235, 1981.
105. Noguchi, T. and Hayashi, S., Peroxisomal and properties of tryptophan aminotransferase in plant leaves, *J. Biol. Chem.*, 255, 2267, 1980.
106. Wightman, F. and Forest, J. C., Properties of plant aminotransferases, *Phytochemistry*, 17, 1455, 1978.
107. Walton, N. J. and Butt, V. S., Glutamate and serine as competing donors for amination of glyoxylate in leaf peroxisomes, *Planta*, 153, 232, 1981.
108. Somerville, C. R. and Ogren, W. L., Photorespiration mutants of *Arabidopsis thaliana* deficient in serine: glyoxylate aminotransferase, *Proc. Natl. Acad. Sci. U.S.A.*, 77, 2684, 1980.
109. Chang, C.-C. and Huang, A. H. C., Metabolism of glycolate in isolated spinach leaf peroxisomes. Kinetics of glyoxylate, oxalate, carbon dioxide, and glycine formation, *Plant Physiol.*, 67, 1003, 1981.
110. Walton, N. J. and Butt, V. S., Metabolism and decarboxylation of glycolate and serine in leaf peroxisomes, *Planta*, 153, 225, 1981.
111. Yang, J. C. and Loewus, F. A., Metabolic conversion of L-ascorbic acid to oxalic acid in oxalate-accumulating plants, *Plant Physiol.*, 56, 283, 1975.
112. Grodzinski, B., A study of formate production and oxidation in leaf peroxisomes during photorespiration, *Plant Physiol.*, 63, 289, 1979.
113. Zelitch, I., Comparison of the effectiveness of glycolic acid and glycine as substrates for photorespiration, *Plant Physiol.*, 50, 109, 1972.
114. Grodzinski, B. and Butt, V. S., Hydrogen peroxide production and the release of carbon dioxide during glycolate oxidation in leaf peroxisomes, *Planta*, 128, 225, 1976.
115. Grodzinski, B., Glyoxylate decarboxylation during photorespiration, *Planta*, 144, 31, 1978.
116. Kisaki, T. and Tolbert, N. E., Glycolate and glyoxylate metabolism by isolated peroxisomes, *Plant Physiol.*, 44, 242, 1969.
117. Halliwell, B. and Butt, V. S., Oxidative decarboxylation of glycolate and glyoxylate by leaf peroxisomes, *Biochem. J.*, 138, 217, 1974.
118. Halliwell, B., Oxidation of formate by peroxisomes and mitochondria from spinach leaves, *Biochem. J.*, 138, 77, 1974.
119. Crosti, P., Intracellular distribution of 10-formyl tetrahydrofolic acid synthetase in spinach leaves, *Ital. J. Biochem.*, 23, 72, 1974.
120. Oliver, D. J., Formate oxidation and oxygen reduction by leaf mitochondria, *Plant Physiol.*, 68, 703, 1981.
121. Ryle, G. J. A. and Hesketh, J. D., Carbon dioxide uptake in nitrogen-deficient plants, *Crop Sci.*, 9, 451, 1969.
122. Zelitch, I. and Gotto, A. M., Properties of a new glyoxylate reductase from leaves, *Biochem. J.*, 84, 541, 1962.
123. Tolbert, N. E., Yamazaki, R. K., and Oeser, A., Localization and properties of hydroxypyruvate and glyoxylate reductases in spinach leaf particles, *J. Biol. Chem.*, 245, 5129, 1970.
124. Zelitch, I., Effect of glycidate, an inhibitor of glycolate synthesis in leaves, on the activity of some enzymes of the glycolate pathway, *Plant Physiol.*, 61, 236, 1978.
125. Peterson, R. B., Enhanced incorporation of tritium into glycolate during photosynthesis by tobacco leaf tissue in the presence of tritiated water, *Plant Physiol.*, 69, 192, 1982.
126. Dench, J. E., Briand, Y., Jackson, C., Hall, D. O., and Moore, A. L., Glycine oxidation in spinach leaf mitochondria, in *Plant Mitochondria*, Ducet, G. and Lance, C., Eds., Elsevier, New York, 1980, 133.
127. Cavalieri, A. J. and Huang, A. H. C., Carrier protein-mediated transport of neutral amino acids into mung bean mitochondria, *Plant Physiol.*, 66, 588, 1980.
128. Somerville, C. R. and Ogren, W. L. Mutants of the cruciferous plant *Arabidopsis thaliana* lacking glycine decarboxylase activity, *Biochem. J.*, 202, 373, 1982.

129. Sarojini, G., Walker, G. H., and Oliver, D. J., Isolation of glycine decarboxylase and implication of a glycine transport mechanism in pea leaf mitochondria, *Plant Physiol.,* 69S, 80, 1982.

130. Day, D. A. and Wiskich, J. T., Glycine transport by pea leaf mitochondria, *FEBS Lett.,* 112, 191, 1980.

131. Bird, I. F., Cornelius, M. J., Keys, A. J., and Whittingham, C. P., Oxidation and phosphorylation associated with the conversion of glycine to serine, *Phytochemistry,* 11, 1587, 1972.

132. Bird, I. F., Cornelius, M. J., Keys, A. J., and Whittingham, C. P., Adenosine triphosphate synthesis and the natural electron acceptor for synthesis of serine from glycine, *Biochem. J.,* 128, 191, 1972.

133. Douce, R., Moore, A. L., and Neuberger, M., Isolation and oxidative properties of intact mitochondria isolated from spinach leaves, *Plant Physiol.,* 60, 625, 1977.

134. Moore, A. L., Jackson, C., Halliwell, B., Dench, J. E., and Hall, D. O., Intramitochondrial localization of glycine decarboxylase in spinach leaves, *Biochem. Biophys. Res. Commun.,* 78, 483, 1977.

135. Moore, A. L., Dench, J. E., Jackson, C., and Hall, D. O., Glycine decarboxylase activity in plant tissue measured by a rapid assay technique, *FEBS Lett.,* 115, 54, 1980.

136. Woo, K. C. and Osmond, C. B., Glycine decarboxylation in mitochondria isolated from spinach leaves, *Aust. J. Plant Physiol.,* 3, 771, 1976.

137. Arron, G. P., Spalding, M. H., and Edwards, G. E., Stoichiometry of carbon-dioxide release and oxygen uptake during glycine oxidation in mitochondria isolated from spinach (*Spinacia oleracea*) leaves, *Biochem. J.,* 184, 457, 1979.

138. Stitt, M., McLilley, R., and Heldt, H. W., Adenine nucleotide levels in the cytosol, chloroplast, and mitochondrion of wheat leaf protoplasts, *Plant Physiol.,* 70, 971, 1982.

139. Journet, E. P., Neuburger, M., and Douce, R., Role of glutamate-oxaloacetate transaminase and malate dehydrogenase in the regeneration of NAD^+ for glycine oxidation by spinach leaf mitochondria, *Plant Physiol.,* 67, 467, 1981.

140. Oliver, D. J., Mechanism of decarboxylation of glycine and glycolate by isolated soybean cells, *Plant Physiol.,* 64, 1048, 1979.

141. Hiraga, K. and Kikuchi, G., The mitochondrial glycine cleavage system. Purification and properties of glycine decarboxylase from chicken liver mitochondria, *J. Biol. Chem.,* 255, 11664, 1980.

142. Hiraga, K. and Kikuchi, G., The mitochondrial glycine cleavage system. Functional association of glycine decarboxylase and amino methyl carrier protein, *J. Biol. Chem.,* 255, 11671, 1980.

143. Kikuchi, G. and Hiraga, K., The mitochondrial glycine cleavage system, *Mol. Cell. Biochem.,* 45, 137, 1982.

144. Woo, K. C., Properties and intramitochondrial localization of serine hydroxymethyl transferase in leaves of higher plants, *Plant Physiol.,* 63, 783, 1979.

145. Wang, D. and Burris, R. H., Carbon metabolism of 14C-labeled amino acids in wheat leaves. II. Serine and its role in glycine metabolism, *Plant Physiol.,* 38, 430, 1963.

146. Clandinin, M. T. and Cossins, E. A., Regulation of mitochondrial glycine decarboxylase from pea mitochondria, *Phytochemistry,* 14, 387, 1975.

147. Gardestrom, P., Bergman, A., and Ericson, I., Inhibition of the conversion of glycine to serine in spinach leaf mitochondria, *Physiol. Plant.,* 53, 439, 1981.

148. Lawyer, A. L. and Zelitch, I., Inhibition of glycine decarboxylation and serine formation in tobacco by glycine hydroxamate and its effect on photorespiratory carbon flow, *Plant Physiol.,* 64, 706, 1979.

149. Usuda, H., Arron, G. D., and Edwards, G. E., Inhibition of glycine decarboxylation by aminoacetonitrile and its effects on photosynthesis in wheat, *J. Exp. Bot.,* 31, 1477, 1980.

150. Créach, E. and Stewart, C. R., Effects of aminoacetonitrile on net photosynthesis, ribulose-1,5-bisphosphate levels, and glycolate pathway intermediates, *Plant Physiol.,* 70, 1444, 1982.

151. Gardeström, P., Bergman, A., and Ericson, I., Oxidation of glycine via the respiratory chain in mitochondria prepared from different parts of spinach, *Plant Physiol.,* 65, 389, 1980.

152. Arron, G. P. and Edwards, G. E., Light-induced development of glycine oxidation by mitochondria from sunflower cotyledons, *Plant Sci. Lett.,* 18, 229, 1980.

153. Neuburger, M. and Douce, R., Oxydation du malate, du NADH et de la glycine par les mitochondries des plantes en C3 et C4, *Physiol. Vég.,* 285, 881, 1977.

154. Miflin, B. J., Marker, A. F. H., and Whittingham, C. P., The metabolism of glycine and glycolate by pea leaves in relation to photosynthesis, *Biochim. Biophys. Acta,* 120, 266, 1966.

155. Somerville, S. C. and Somerville, C. R., Effect of carbon dioxide on photorespiratory flux determined from glycine accumulation in a mutant of *Arabidopsis thaliana, J. Exp. Bot.,* 34, 415, 1983.

156. Heber, U., Kirk, M. R., Gimmler, H., and Schafer, G., Uptake and reduction of glycerate by isolated chloroplasts, *Planta,* 120, 31, 1974.

157. Bird, I. F., Cornelius, M. J., Keys, A. J., and Whittingham, C. P., Intracellular site of sucrose synthesis in leaves, *Phytochemistry,* 13, 59, 1974.

158. Usuda, H. and Edwards, G. E., Localization of glycerate kinase and some enzymes for sucrose synthesis in C3 and C4 plants, *Plant Physiol.,* 65, 1017, 1980.

159. Robinson, S. P., Light stimulates glycerate uptake by spinach chloroplasts, *Biochem. Biophys. Res. Commun.*, 106, 1027, 1982.

160. Robinson, S. P., Transport of glycerate across the envelope membrane of isolated spinach chloroplasts, *Plant Physiol.*, 70, 1032, 1982.

161. Rabson, R., Tolbert, N. E., and Kearney, P. C., Formation of serine and glyceric acid by the glycolate pathway, *Arch. Biochem. Biophys.*, 98, 154, 1962.

162. Tolbert, N. E., Glycolate metabolism by higher plants and algae, in *Encyclopedia of Plant Physiology, New Series,* Vol. 6, Gibbs, M. and Latzko, E., Eds., Springer-Verlag, New York, 1979, 338.

163. Mulligan, R. M. and Tolbert, N. E., Properties of a membrane-bound phosphatase from the thylakoids of spinach chloroplasts, *Plant Physiol.*, 66, 1169, 1980.

164. Robinson, S. P., 3-Phosphoglycerate phosphatase activity in chloroplast preparations as a result of contamination by acid phosphatase, *Plant Physiol.*, 70, 645, 1982.

165. Mahon, J. D., Fock, H., and Canvin, D. T., Changes in specific radioactivity of sunflower leaf metabolites during photosynthesis in $^{14}CO_2$ and $^{12}CO_2$ at three concentrations of CO_2, *Planta*, 120, 245, 1974.

166. Larsson, C. and Albertsson, E., Enzymes related to serine synthesis in spinach chloroplasts, *Physiol. Plant.*, 45, 7, 1979.

167. Keys, A. J., Bird, I. F., Cornelius, M. J., Lea, P. J., Wallsgrove, R. M., and Miflin, B. J., Photorespiratory nitrogen cycle, *Nature (London)*, 275, 741, 1978.

168. Wallsgrove, R. M., Keys, A. J., Bird, I. F., Cornelius, M. J., Lea, D. J., and Miflin, B. J., The location of glutamine synthetase in leaf cells and its role in the reassimilation of ammonia released in photorespiration, *J. Exp. Bot.*, 31, 1005, 1980.

169. Mann, A. F., Fenten, P. A., and Stewart, G. R., Tissue localization of barley (*Hordeum vulgare*) glutamine synthetase isoenzymes, *FEBS Lett.*, 110, 265, 1980.

170. Somerville, S. C. and Ogren, W. L., *Arabidopsis thaliana* mutant defective in chloroplast dicarboxylate transport, *Proc. Natl. Acad. Sci. U.S.A.*, 80, 1290, 1983.

171. Somerville, C. R. and Ogren, W. L., Inhibition of photosynthesis in mutants of *Arabidopsis* lacking glutamate synthase activity, *Nature (London)*, 286, 257, 1980.

172. Canvin, D. T., Lloyd, N. D. H., Fock, H., and Przybylla, K., Glycine and serine metabolism and photorespiration, in *CO₂ Metabolism and Plant Productivity,* Burris, R. H. and Black, C. C., Eds., University Park Press, Baltimore, 1976, 161.

173. Agrawal, P. K. and Fock, H., The specific radioactivity of glycolic acid in relation to the specific activity of carbon dioxide evolved in light in photosynthesizing sunflower leaves, *Planta*, 138, 257, 1978.

174. Oliver, D. J., Role of glycine and glyoxylate decarboxylatory in photorespiratory CO_2 release, *Plant Physiol.*, 68, 1031, 1981.

175. Zelitch, I., Photorespiration: studies with whole tissues, in *Enclyclopedia of Plant Physiology, New Series,* Vol. 6, Gibbs, M. and Latzko, E., Eds., Springer-Verlag, New York, 1979, 353.

176. Lloyd, N. D. H. and Canvin, D. T., Photosynthesis and photorespiration in sunflower selections, *Can. J. Bot.*, 55, 3006, 1977.

177. Gerbaud, A. and André, M., Photosynthesis and photorespiration in whole plants of wheat, *Plant Physiol.*, 64, 735, 1979.

178. Kumarasinghe, K. S., Keys, A. J., and Whittingham, C. P., The flux of carbon through the glycolate pathway during photosynthesis by wheat leaves, *J. Exp. Bot.*, 28, 1247, 1977.

179. Servaites, J. C., Schrader, L. E., and Edwards, G. E., Glycolate synthesis in a C_3, C_4 and intermediate photosynthetic plant type, *Plant Cell Physiol.*, 19, 1399, 1978.

180. Rathnam, C. K. M. and Chollet, R., Photosynthetic and photorespiratory carbon metabolism in mesophyll protoplasts and chloroplasts isolated from isogenic diploid and tetraploid cultivars of ryegrass (*Lolium perenne* L.) *Plant Physiol.*, 65, 489, 1980.

181. Hitz, W. D. and Stewart, C. R., Oxygen and carbon dioxide effects on the pool size of some photosynthetic and photorespiratory intermediates in Soybean (*Glycine max* [L.] Merr.), *Plant Physiol.*, 65, 442, 1980.

182. Jensen, R. G. and Bahr, J. T., Ribulose 1,5-bisphosphate carboxylase-oxygenase, *Annu. Rev. Plant. Physiol.*, 28, 379, 1977.

183. Oliver, D. J., The interaction between O_2 and CO_2 concentrations on the regulation of glycolate synthesis in tobacco leaf discs, *Plant Sci. Lett.*, 15, 35, 1979.

184. Zelitch, I., Increased rate of net photosynthetic carbon dioxide uptake caused by the inhibition of glycolate oxidase, *Plant Physiol.*, 41, 1623, 1966.

185. Somerville, C. R. and Ogren, W. L., Genetic modification of photorespiration, *Trends Biochem. Sci.*, 7, 171, 1982.

186. Gunther, G., Baumann, G., Klos, J., and Bafanz, J., Effects of inhibitors and herbicides on net photosynthesis and photorespiration of sugar beet (*Beta vulgaris*) and white lambsquarter (*Chenopodium album*), *Biochem. Physiol. Pflanz.*, 174, 616, 1979.

187. Cook, C. M. and Tolbert, N. E., Inhibition of spinach ribulose bisphosphate carboxylase/oxygenase by glyoxylate, *Plant Physiol.*, 69S, 52, 1982.
188. Lawyer, A. L. and Zelitch, I., Inhibition of glutamate: glyoxylate aminotransferase activity in tobacco leaves and callus by glycidate, an inhibitor of photorespiration, *Plant Physiol.*, 61, 242, 1978.
189. Chollet, R., Effect of glycidate on glycolate formation and photosynthesis in isolated chloroplasts, *Plant Physiol.*, 57, 237, 1976.
190. Wildner, G. F. and Larsson, C., Effects of glycidate on carbon dioxide fixation with isolated spinach chloroplasts, *Plant Physiol.*, 63, 887, 1979.
191. Oliver, D. J. and Zelitch, I., Increasing photosynthesis by inhibiting photorespiration with glyoxylate, *Science*, 196, 1450, 1977.
192. Oliver, D. J., Effect of glyoxylate on the sensitivity of net photosynthesis to oxygen (the Warburg effect) in tobacco, *Plant Physiol.*, 62, 938, 1978.
193. Oliver, D. J., The effect of glyoxylate on photosynthesis and photorespiration by isolated soybean mesophyll cells, *Plant Physiol.*, 65, 888, 1980.
194. Oliver, D. J. and Zelitch, I., Metabolic regulation of glycolate synthesis, photorespiration and net photosynthesis in tobacco by L-glutamate, *Plant Physiol.*, 59, 688, 1977.
195. Baumann, G. and Gunther, G., Effects of glyoxylate, glycine and serine on the assimilation of $^{14}CO_2$ into mesophyll cells of *Chenopodium album*, *Biochem. Physiol. Pflanz.*, 174, 160, 1979.
196. Yun, S. J., Ishi, R., Hycon, S. B., Suzuki, A., Murata, Y., and Tamura, S., Effects of some chemicals on photorespiration and photosynthesis in the excised rice leaves, *Agric. Biol. Chem.*, 43, 2207, 1979.
197. Ruffo, A., Testa, E., Adinolfig, A., Pelizza, G., and Moratti, R., Control of the citric acid cycle by glyoxylate. Mechanism of the inhibition by oxalomalate and γ-hydroxy-α-oxoglutarate, *Biochem. J.*, 103, 19, 1967.
198. Peterson, R. B., Regulation of glycine decarboxylase and L-serine hydroxymethyl transferase by glyoxylate in tobacco leaf mitochondrial preparations, *Plant Physiol.*, 70, 61, 1982.
199. Akazawa, T., Ribulose-1,5-bisphosphate carboxylase, in *Encyclopedia of Plant Physiology, New Series*, Vol. 6, Gibbs, M. and Latzko, E., Eds., Springer-Verlag, New York, 1979.
200. Ku, M. S. B., Schmitt, M. R., and Edwards, G. E., Quantitative determination of RuBP carboxylase-oxygenase protein in leaves of several C_3 and C_4 plants, *J. Exp. Bot.*, 30, 89, 1979.
201. Berhow, M. A., Saluja, A., and McFadden, B. A., Rapid purification of D-ribulose 1,5-bisphosphate carboxylase by vertical sedimentation in a reoriented gradient, *Plant Sci. Lett.*, 27, 51, 1982.
202. Lorimer, G. H., Badger, M. R., and Andrews, T. J., D-ribulose-1,5-bisphosphate carboxylase-oxygenase. Improved methods for activation and assay of catalytic activities, *Anal. Biochem.*, 78, 66, 1977.
203. Bird, I. F., Cornelius, M. J., and Keys, A. J., Effect of carbonic anhydrase on the activity of ribulose bisphosphate carboxylase, *J. Exp. Bot.*, 31, 365, 1980.
204. Pierce, J. W., McCurry, S. D., Mulligan, R. M., and Tolbert, N. E., Activation and assay of ribulose-1,5-bisphosphate carboxylase/oxygenase, *Methods Enzymol.*, 89, 47, 1982.
205. Andrews, T. J. and Abel, K. M., Kinetics and subunit interactions of ribulose bisphosphate carboxylase-oxygenase from the cyanobacterium *Synechoccus* sp., *J. Biol. Chem.*, 256, 8445, 1981.
206. McNiel, P. H., Foyer, C. H., Walker, D. A., Bird, I. F., Cornelius, M. J., and Keys, A. J., Similarity of ribulose-1,5-bisphosphate carboxylases of isogenic diploid and tetraploid ryegrass (*Lolium perenne L.*) cultivars, *Plant Physiol.*, 67, 530, 1981.
207. Kent, S. S. and Young, J. D., Simultaneous kinetic analysis of ribulose-1,5-bisphosphate carboxylase/oxygenase activities, *Plant Physiol.*, 65, 465, 1980.
208. Jordan, D. B. and Ogren, W. L., A sensitive assay procedure for simultaneous determination of ribulose 1,5-bisphosphate carboxylase and oxygenase activities, *Plant Physiol.*, 67, 237, 1981.
209. Chen, K., Gray, J. C., and Wildman, S. G., Fraction I protein and the origin of polyploid wheats, *Science*, 190, 1304, 1975.
210. Kung, S. D., Tobacco fraction I protein: a unique genetic marker, *Science*, 191, 429, 1976.
211. McIntosh, L., Poulsen, C., and Bogorad, L., Chloroplast gene sequence for the large subunit of ribulose bisphosphate carboxylase of maize, *Nature (London)*, 288, 556, 1980.
212. Zurawski, G., Perrot, B., Bottomley, W., and Whitfield, P. R., The structure of the gene for the large subunit of ribulose-1,5-bisphosphate carboxylase from spinach chloroplast DNA, *Nucleic Acid Res.*, 9, 3251, 1981.
213. Highfield, P. E. and Ellis, R. J., Synthesis and transport of the small subunit of chloroplast ribulose bisphosphate carboxylase, *Nature (London)*, 271, 420, 1978.
214. Chua, N. H. and Schmidt, G. W., Post-translated transport into intact chloroplasts of a precursor to the small subunit of ribulose-1,5-bisphosphate carboxylase, *Proc. Natl. Acad. Sci. U.S.A.*, 75, 6110, 1978.
215. Roy, H., Bloom, M., Milos, P., and Monroe, M., Studies on the assembly of large subunits of ribulose bisphosphate carboxylase in isolated pea chloroplasts, *J. Cell Biol.*, 94, 20, 1982.

216. Bedbrook, J. R., Smith, S. M. and Ellis, R. J., Molecular cloning and sequencing of cDNA encoding the precursor to the small subunit of chloroplast ribulose-1,5-bisphosphate carboxylase, *Nature (London)*, 287, 629, 1980.

217. Broglie, R., Bellemare, G., Bartlett, S. G., Chua, N. H., and Cashmore, A. R., Cloned DNA sequences complementary to mRNAs encoding precursors to the small subunit of ribulose-1,5-bisphosphate carboxylase and a chlorophyll a/b binding polypeptide, *Proc. Natl. Acad. Sci. U.S.A.*, 78, 7304, 1981.

218. Lorimer, G. H., Badger, M. R., and Andrews, J. T., The activation of ribulose-1,5-bisphosphate carboxylase by carbon dioxide and magnesium ions. Equilibria, kinetics, a suggested mechanism and physiological implications, *Biochemistry*, 15, 529, 1976.

219. Wildner, G. F. and Henkel, J., Differential reactivation of ribulose-1,5-bisphosphate oxygenase with low carboxylase activity by Mn^{2+}, *FEBS Lett.*, 91, 99, 1978.

220. Christeller, J. T. and Laing, W. A., Effects of manganese ions and magnesium ions on the activity of soya-bean ribulose bisphosphate carboxylase/oxygenase, *Biochem. J.*, 183, 747, 1979.

221. Martin, M. N. and Tabita, F. R., Differences in the kinetic properties of the carboxylase and oxygenase activities of ribulose bisphosphate carboxylase/oxygenase, *FEBS Lett.*, 129, 39, 1981.

222. Bhagwat, A. S., Ramakrishna, J., and Sane, P. V., Specific inhibition of oxygenase activity of ribulose-1,5-diphosphate carboxylase by hydroxylamine, *Biochem. Biophys. Res. Commun.*, 83, 954, 1978.

223. Bhagwat, A. S. and Sane, P. V., Evidence for the involvement of superoxide anions in the oxygenase reaction of ribulose-5-diphosphate carboxylase, *Biochem. Biophys. Res. Commun.*, 84, 865, 1978.

224. Okabe, K. I., Codd, G. A., and Stewart, W. D. P., Hydroxylamine stimulates carboxylase activity and inhibits oxygenase activity of cyanobacterial RuBP carboxylase/oxygenase, *Nature (London)*, 279, 525, 1979.

225. Wildner, G. F. and Henkel, J., Specific inhibition of the oxygenase activity of ribulose-1,5-bisphosphate carboxylase, *Biochem. Biophys. Res. Commun.*, 69, 268, 1976.

226. Wildner, G. F. and Henkel, J., Preservation of RuBP carboxylase without oxygenase activity during anaerobiosis, *FEBS Lett.*, 113, 81, 1980.

227. Andrews, J. T. and Lorimer, G. H., Photorespiration — still unavoidable?, *FEBS Lett.*, 90, 1, 1978.

228. Yeoh, H. H., Badger, M. R., and Watson, L., Variations in $Km (CO_2)$ of ribulose-1,5-bisphosphate carboxylase among grasses, *Plant Physiol.*, 66, 1110, 1980.

229. Yeoh, H. H., Badger, M. R., and Watson, L., Variations in kinetic properties of ribulose-1,5-bisphosphate carboxylases among plants, *Plant Physiol.*, 67, 1151, 1981.

230. Jordan, D. B. and Ogren, W. L., Species variation in the specificity of ribulose bisphosphate carboxylase/oxygenase, *Nature (London)*, 291, 513, 1981.

231. Bird, I. F., Cornelius, M. J., and Keys, A. J., Affinity of RuBP carboxylases for carbon dioxide and inhibition of the enzymes by oxygen, *J. Exp. Bot.*, 33, 1004, 1982.

232. Garret, M. K., Control of photorespiration at RuBP carboxylase/oxygenase level in ryegrass cultivars, *Nature (London)*, 274, 913, 1978.

233. Christeller, J. T., The effects of bivalent cations on ribulose bisphosphate carboxylase/oxygenase, *Biochem. J.*, 193, 839, 1981.

234. Rejda, J. M., Johal., S., Chollet, R., Enzymic and physiochemical characterization of ribulose-1,5-bisphosphate carboxylase/oxygenase from diploid and tetraploid cultivars of perennial ryegrass, *Arch. Biochem. Biophys.*, 210, 617, 1981.

235. Bravdo, B., Palgi, A., Lurie, S., and Frenkel, C., Changing ribulose diphosphate carboxylase/oxygenase activity in ripening tomato fruit, *Plant Physiol.*, 60, 309, 1977.

236. Martin, B. A., Gauger, J. A., and Tolbert, N. E., Changes in activity of ribulose-1,5-bisphosphate carboxylase/oxygenase and three peroxisomal enzymes during tomato fruit development and ripening, *Plant Physiol.*, 63, 486, 1979.

237. Okabe, K., Properties of ribulose disphosphate carboxylase/oxygenase in the tobacco aurea mutant *Su/su* Var. *Aurea*, *Z. Naturforsch.*, 31, 781, 1977.

238. Zelitch, I. and Day, P. R., Variation in photorespiration. The effect of genetic differences in photorespiration on net photosynthesis in tobacco, *Plant Physiol.*, 43, 1838, 1968.

239. Kung, S. D. and Marsho, T. V., Regulation of RuBP carboxylase/oxygenase activity and its relationship to plant photorespiration, *Nature (London)*, 259, 325, 1976.

240. Koivuniemi, P. J., Tolbert, N. E., and Carlson, P. S., Ribulose-1,5-bisphosphate carboxylase/oxygenase and polyphenol oxidase in the tobacco mutant *Su/su* and three green revertant plants, *Plant Physiol.*, 65, 828, 1980.

241. Hirai, A. and Tsunewaki, K., Genetic diversity of the cytoplasm in *Triticum* and *Aegilops*. VIII. Fraction I protein of 39 cytoplasms, *Genetics*, 99, 487, 1981.

242. Steer, M. W. and Kernoghan, D., Nuclear and cytoplasm genome relationships in the genus *Avena*: analysis by isoelectric focusing of ribulose bisphosphate carboxylase subunits, *Biochem. Genet.*, 15, 273, 1977.

243. Gatenby, A. A. and Cocking, E. C., Polypeptide composition of fraction I protein subunits in the genus *Petunia, Plant Sci. Lett.,* 10, 97, 1977.
244. Uchimiya, H. and Wildman, S. G., Evolution of fraction I protein in relation to origin of amphidiploid *Brassica* species and other members of the Cruciferae, *J. Hered.,* 69, 299, 1978.
245. Murphy, T., Immunochemical comparisons of ribulose bisphosphate carboxylases using antisera to tobacco and spinach enzymes, *Phytochemistry,* 17, 439, 1978.
246. Poulsen, C., Comments on the structure and function of the large subunit of the enzyme ribulose bisphosphate carboxylase oxygenase, *Carlsberg Res. Commun.,* 46, 259, 1981.
247. Spreitzer, R. J. and Mets, L. J., Non-mendelian mutation affecting ribulose-1,5-bisphosphate carboxylase structure and activity, *Nature (London),* 285, 114, 1980.
248. Andersen, K., Mutations altering the catalytic activity of a plant-type ribulose bisphosphate carboxylase/oxygenase in *Alcaligenes eutrophus, Biochim. Biophys. Acta,* 585, 1, 1979.
249. Martin, P. G., Amino acid sequence of the small subunit of ribulose-1,5-bisphosphate carboxylase from spinach, *Aust. J. Plant Physiol.,* 6, 401, 1979.
250. Takruri, I. A. H., Boulter, D., and Ellis, R. J., Amino acid sequence of the small subunit of ribulose-1,5-bisphosphate carboxylase of *Pisum sativum, Phytochemistry,* 20, 413, 1981.
251. Haslett, B. G., Yarwood, A., Evans, I. M., and Boulter, D., Studies on the small subunit of fraction I protein from *Pisum sativum, Photochemistry,* 20, 413, 1981.
252. Strobaek, S., Gibbons, G. C., Haslett, B. G., Boulter, D., and Wildman, S. G., On the nature of the polymorphism of the small subunit of ribulose-1,5-diphosphate carboxylase in the amphidiploid *Nicotiana tabacum, Carlsberg Res. Commun.,* 41, 335, 1976.
253. Rhodes, P. R., Kung, S. D., and Marsho, T. V., Relationship of ribulose-1,5-bisphosphate carboxylase-oxygenase specific activity to subunit composition, *Plant Physiol.,* 65, 69, 1980.
254. Menz, K. M., Moss, D. N., Connell, R. Q., and Brun, W. A., Screening for photosynthetic efficiency, *Crop Sci.,* 9, 692, 1969.
255. Widholm, J. M. and Ogren, W. L., Photorespiratory-induced senescence of plants under conditions of low carbon dioxide, *Proc. Natl. Acad. Sci. U.S.A.,* 63, 668, 1969.
256. Connell, R. Q., Brun, W. A., and Moss, D. N., A search for high net photosynthetic rate among soybean genotypes, *Crop Sci.,* 9, 840, 1969.
257. Sharma, R. K., Griffing, B., and Scholl, R. L., Variations among races of *Arabidopsis thaliana* (L.) Heynh. for survival in limited carbon dioxide, *Theor. Appl. Genet.,* 54, 11, 1979.
258. Nelson, C. J., Assay, K. H., and Patton, L. D., Photosynthetic responses of tall fescue to selection for longevity below the CO_2 compensation point, *Crop Sci.,* 15, 629, 1975.
259. Berlyn, M. B., A mutational approach to the study of photorespiration, in *Photosynthetic Carbon Assimilation,* Siegelman, H. W. and Hind, G., Eds., Plenum Press, New York, 1978, 153.
260. Berlyn, M. B., Isolation and characterization of isonicotinic acid hydrazide resistant mutants of *Nicotiana tabacum, Theor. Appl. Genet.,* 58, 19, 1980.
261. Zelitch, I. and Berlyn, M. B., Altered glycine decarboxylation inhibition in isonicotinic acid hydrazide-resistant mutant callus lines and in regenerated plants and seed progeny, *Plant Physiol.,* 69, 198, 1982.
262. Lawyer, A. L., Berlyn, M. B., and Zelitch, I., Isolation and characterization of glycine hydroxamate-resistant cell lines of *Nicotiana tabaccum, Plant Physiol.,* 66, 334, 1980.
263. Boynton, J., Nobs, M. A., Björkman, O., and Pearcy, R. W., Hybrids between *Atriplex* species with and without β-carboxylation photosynthesis, *Carnegie Inst. Washington Yearb.,* 69, 629, 1970.
264. Nobs, M. A., Hybridization in *Atriplex, Carnegie Inst. Washington Yearb.,* 75, 421, 1976.
265. Osmond, C. B., Björkman, O., and Anderson, D. J., *Physiological Processes in Plant Ecology. Toward a Synthesis with Atriplex,* Springer-Verlag, New York, 1980.
266. Pearcy, R. W. and Björkman, O., Hybrids between *Atriplex* species with and without β-carboxylation photosynthesis. Biochemical characteristics, *Carnegie Inst. Washington Yearb.,* 69, 632, 1970.
267. Hattersley, P. W., Watson, L., and Osmond, C. B., *In situ* immunofluorescent labeling of ribulose-1,5-bisphosphate carboxylase in leaves of C_3 and C_4 plants, *Aust. J. Plant Physiol.,* 4, 523, 1977.
268. Brown, R. H. and Brown, W. V., Photosynthetic characteristics of *Panicum milioides,* a species with reduced photorespiration, *Crop Sci.,* 15, 681, 1
269. Morgan, J. A. and Brown, R. H., Photosynthesis in grass species differing in carbon dioxide fixation pathways. II. A search for species with intermediate gas exchange and anatomical characteristics, *Plant Physiol.,* 64, 257, 1979.
270. Kanai, R. and Kashiwagi, M., *Panicum milioides,* a Gramineae plant having Kranz anatomy without C_4 photosynthesis, *Plant Cell Physiol.,* 16, 669, 1975.
271. Brown, R. H., Bouton, J. H., Rigsby, L., and Rigler, M., Photosynthesis of grass species differing in carbon dioxide fixation pathways. VIII. Ultrastructural characteristics of *Panicum* species in the Laxa group, *Plant Physiol.,* 71, 425, 1983.
272. Keck, R. W. and Ogren, W. L., Differential oxygen response of photosynthesis in soybean and *Panicum milioides, Plant Physiol.,* 58, 552, 1976.

273. Brown, R. H. and Morgan, J. A., Photosynthesis of grass species differing in carbon dioxide fixation pathways. VI. Differential effects of temperature and light intensity on photorespiration in C_3, C_4 and intermediate species, *Plant Physiol.*, 66, 541, 1980.

274. Rathnam, C. K. M. and Chollet, R., Photosynthetic carbon metabolism in *Panicum milioides*, a C_3-C_4 intermediate species: evidence for a limited C_4 dicarboxylic acid pathway of photosynthesis, *Biochim. Biophys. Acta*, 548, 500, 1979.

275. Holaday, A. S., Harrison, A. T., and Chollet, R., Photosynthetic/photorespiratory CO_2 exchange characteristics of the C_3-C_4 intermediate species, *Moricandia arvensis, Plant Sci. Lett.*, 27, 181, 1982.

276. Ku, S. B., Edwards, G. E., and Kanai, R., Distribution of enzymes related to C_3 and C_4 pathway of photosynthesis between mesophyll and bundle sheath cells of *Panicum hians* and *Panicum milioides*, *Plant Cell Physiol.*, 17, 615, 1976.

277. Goldstein, L. D., Ray, T. B., Kestler, D. P., Mayne, B. C., Brown, R. H., and Black, C. C., Biochemical characterization of *Panicum* species which are intermediate between C_3 and C_4 photosynthesis plants, *Plant Sci. Lett.*, 6, 85, 1976.

278. Edwards, G. E., Ku, M. S. B., and Hatch, M. D., Photosynthesis in *Panicum milioides* a species with reduced photorespiration, *Plant Cell Physiol.*, 23, 1185, 1982.

279. Winter, K., Holtum, J. A. M., Edwards, G. E., and O'Leary, M. H., Effect of low relative humidity on $d^{13}C$ value in two C_3 grasses and in *Panicum milioides*, a C_3-C_4 intermediate species, *J. Exp. Bot.*, 33, 88, 1982.

280. Ku, S. B. and Edwards, G. E., Photosynthetic efficiency of *Panicum hians* and *Panicum milioides* in relation to C_3 and C_4 plants, *Plant Cell Physiol.*, 19, 665, 1978.

281. Morgan, J. A., Brown, R. H., and Reger, B. J., Photosynthesis in grass species differing in carbon dioxide fixation pathways. III. Oxygen response and enzyme activities of species in the *Laxa* group of *Panicum, Plant Physiol.*, 65, 156, 1980.

282. Bouton, J. H., Brown, R. H., Bolton, J. K., and Campagnoli, R. P., Photosynthesis of grass species differing in carbon dioxide fixation pathways. VIII. Chromosome numbers, metaphase I chromosome behavior, and mode of reproduction of photosynthetically distinct *Panicum* species, *Plant Physiol.*, 67, 433, 1981.

283. Teeri, J. and Stowe, L. G., Climate patterns and the distribution of C_4 grasses in North America, *Oecologia*, 23, 1, 1976.

284. Ehleringer, J. and Björkman, O., Quantum yields for CO_2 uptake in C_3 and C_4 plants. Dependence on temperature, CO_2, and O_2 concentration, *Plant Physiol.*, 59, 86, 1977.

285. Gould, S. J. and Lewontin, R. C., The spandrels of San Marco and the Panglossian paradigm: a critique of the adaptionist programme, *Proc. R. Soc. Lond. B.*, 205, 581, 1979.

286. Osmond, C. B. and Björkman, O., Simultaneous measurements of oxygen effects on net photosynthesis and glycolate metabolism in C_3 and C_4 species of *Atriplex, Carnegie Inst. Washington Yearb.*, 71, 141, 1972.

287. Cornic, G., Effet exercé sur l'activité photosynthétique du *Sinapsis alba* L. par une inhibition temporaire de la photorespiration se déroulant dans un air sans CO_2, *C. R. Acad. Sci. Paris*, 282, 1955, 1976.

288. Cornic, G., La photorespiration se déroulant dans un air sans CO_2, a-t-elle une fonction?, *Can. J. Bot.*, 56, 2128, 1978.

289. Cornic, G., Woo, K. C., and Osmond, C. B., Photoinhibition of CO_2-dependent O_2 evolution by intact chloroplasts isolated from spinach leaves, *Plant Physiol.*, 70, 1310, 1982.

290. Powles, S. B. and Osmond, C. B., Inhibition of the capacity and efficiency of photosynthesis in bean leaflets illuminated in a CO_2-free atmosphere at low oxygen: a possible role for photorespiration, *Aust. J. Plant Physiol.*, 5, 619, 1978.

291. Powles, S. B., Osmond, C. B., and Thorne, S. W., Photoinhibition of intact attached leaves of C_3 plants illuminated in the absence of both carbon dioxide and photorespiration, *Plant Physiol.*, 64, 982, 1979.

292. Powles, S. B. and Critchley, C., Effect of light intensity during growth on photoinhibition of intact attached bean leaves, *Plant Physiol.*, 65, 1181, 1980.

293. Lorimer, G. H., personal communication.

Chapter 8

SYNTHESIS AND BREAKDOWN OF STARCH*

Charles D. Boyer

TABLE OF CONTENTS

* Contribution No. 19, Department of Horticulture, The Pennsylvania State University. Authorized for publication as paper No. 6585 in the journal series of the Pennsylvania Agricultural Experiment Station.

I. INTRODUCTION

Starch is found in all organs of most higher plants including seeds, stems, leaves, roots, fruits, and pollen.[1] As the principal storage carbohydrate of plants, starch is stored as insoluble granules. The solubilization and fractionation of starch granules yields two populations of molecules. Amylose is primarily a linear molecule of glucose units linked α-1,4. However, amylose may contain an occasional branch point (α-1,6 linkage), approximately one for every 1000 glucose residues.[2,3] Amylopectin is a highly branched glucose polymer and contains an α-1,6 branch point for every 20 to 26 α-1,4 linkages. Amylose generally has a degree of polymerization of 600 to 3000 glucose units and a weighted average molecular weight of approximately 10^6.[2,4] Amylopectin molecules are larger and have a degree of polymerization of 6,000 to 60,000 glucose units (weighted average molecular weight of 10^8).[2,4] Starch granules from higher plants contain mixtures of amylose and amylopectin, generally in a 1:3 ratio, although starch granule morphology varies widely among species.

Starch in higher plants is found in cellular organelles, the plastids. However, starch granules in the red algae are found outside the chloroplasts while starch granules of green algae are formed within the chloroplast.[1] Green algae starch granules contain both amylose and amylopectin.[5] The starch from red algae contains only amylopectin[6] or a polysaccharide intermediate to amylopectin and the highly branched glycogens.[7] The prokaryotes, bacteria and blue green algae, contain no plastids and store carbohydrate as soluble glycogen.[8] Therefore, two changes in the storage of energy as carbohydrate appear to have occurred during evolution. The most obvious specialization is the cellular compartmentation of starch metabolism in plastids. Equally significant, however, is the trend toward more linear polymers, i.e., amylose > amylopectin > glycogen, and the formation of insoluble granules.

Although the starches from higher plants are compositionally similar, starches can be divided into two types, transient or transitory (regularly metabolized) and stable with extended existence as found in seeds or roots. The most familiar transient starch is found in leaves, and varies with the diurnal variation in photosynthesis. Based on the estimated levels of starch synthesis and degradation as well as net accumulation of starch at inorganic phosphate levels found in photosynthetically active leaves, starch synthesis and degradation may be simultaneous events.[9] Thus, net accumulation would result under conditions where synthesis rates exceed degradation rates. Transitory starch can also be found as a component in the development of some seeds (in particular oil seeds). For example, developing soybeans accumulate up to 10% starch on a dry weight basis during the first 38 days after anthesis (mid development).[10] By maturation, starch has been mobilized when the level has fallen to less than 1% of the dry weight and very few plastids contain starch. During germination, starch reappears and accumulates during the first 5 days after imbibition. After 5 days, the starch is rapidly degraded.[10]

Whether starch is transitory or stable, starch is the primary reserve carbohydrate in higher plants and therefore is a critical component of normal growth and development. The economic importance of starch, however, is difficult to fully estimate. Cereals are clearly valued for their starchy endosperms, and a large portion of total yield for these crops is starch. Developing nations often can not afford to convert agricultural production from cereals to high protein crops since the loss of yield in caloric terms would result in increased hunger and malnutrition.[11] As most protein in a calorie-deficient diet is used to meet the caloric needs of an individual, protein becomes an expensive source of calories.[12] In industrialized nations, the uses of starch (or modified starch) in foods or as a raw product for the food industry (i.e., for the production of high fructose corn sweeteners) represents over one billion dollars in sales annually.

Starch biosynthesis and degradation have been regular topics of review. Reviews appearing in the last 5 years have dealt with the biochemistry and regulation of starch synthesis,[13-15] the genetic control of starch synthesis,[16-17] and the use of starch mutants for improving sweet corn cultivars[18] or for industrial purposes.[19] In addition, a number of books are available on starch.[20-22] Obviously, no review can cover all aspects of starch metabolism. Therefore, this review will describe and discuss the metabolism of starch as related to the objectives of plant breeding. In order to facilitate the development of the review, this chapter has been divided into three sections. The first section will describe the pathways of starch metabolism. Particular emphasis is placed on cellular compartmentation and proposed points or enzymatic reactions with regulatory properties. Genetic mutations of starch metabolism will be discussed in the second section. Although most mutations are known for cereal endosperms, evidence for homologous mutations will be presented. The use of these starch mutants as the raw material for plant breeding will be considered. The final section will look at evidence for quantitative variation in starch metabolism and the implication of this variation for plant breeding.

II. BIOCHEMICAL CONSIDERATIONS

Enzymatic reactions of starch metabolism have been known for years. Many of the reactions were among the first enzymatic activities described. The discovery of glucose transfer from UDP-glucose to starch was made in 1961,[23] while other reactions were known previously. Although the enzymatic reactions are now well known and characterized and pathways outlined, little is known about the process(es) involved in the initiation and formation of starch granule structure. Similarly, the pathway of starch degradation can be described as enzymatic reactions, but the sequential steps remain to be further defined. Because the molecular structure of starch is variable (both amylose and amylopectin fractions are populations of molecules of varying structure), no "primary" structure exists for starch. This lack of a clearly defined structure may prevent the description of starch metabolism in precise sequential steps. In spite of this limitation, progress has been made in defining the regulation of starch metabolism. Three types of information are important: (1) the cellular compartmentation of starch metabolism, (2) the enzymatic reactions, and (3) points and reactions with regulatory properties.

A. Compartmentation of Starch Metabolism

Starch biosynthesis in higher plants occurs in plastids. Chloroplasts in green tissue will accumulate starch during the day and mobilize all or some of this starch during subsequent dark periods. Through the use of isolated chloroplasts, a number of enzymes including enzymes of gluconeogenesis and sugar interconversions have been demonstrated to be chloroplastic.[24] Cytosolic forms of these enzymes are also present. ADP-glucose pyrophosphorylase was found to be exclusively located in the chloroplasts of spinach leaves.[25] A second detailed study of spinach leaves demonstrated that all of the starch synthase, starch branching enzyme activities in addition to ADP-glucose pyrophosphorylase activity were located in the chloroplasts.[26] However, UDP-glucose pyrophosphorylase and sucrose-6-P synthase are found only in the cytosol.[27]

The isolation of the amyloplast (leucoplast) from storage tissues has proven more difficult. Electron microscopy has demonstrated the initial formation of starch granules within amyloplasts and the continuity of amyloplast membranes during the early development of cereal endosperms,[20,28,29] potato tubers,[30] and potato roots.[31] As the tissue develops and starch granules grow in size, the tissues become difficult to section due to the high concentration of starch granules. Therefore, in older tissues, mem-

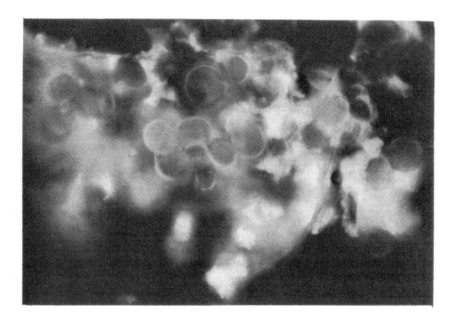

FIGURE 1.   Maize kernel slices treated with fluorescent-labeled antibodies prepared against maize endosperm starch branching enzyme IIa. Note fluorescence associated with amyloplasts. Similar results were obtained with endosperm starch branching enzymes I and IIb.

branes often appear to be discontinuous,[32] but continuous membranes have been observed.[30] The starch granule also restricts the use of aqueous isolation of amyloplasts since the starch granule ruptures the membranes during isolation. The general consensus is that starch synthesis is compartmented in amyloplasts in storage tissues in spite of the paucity of evidence.

Recently, nonaqueous procedures have been developed for the isolation of amyloplasts from maize endosperm.[33] Although enzyme activities were inhibited by the organic solvents used during the isolation, the levels of metabolites were estimated.[34] Based on the levels of neutral sugars, inorganic phosphate, intermediates of gluconeogenesis, and organic acids, Shannon[35] proposed that the amyloplast and chloroplasts were homologous and contain many of the same enzymes. However, the identification of sucrose in the amyloplast preparation suggested that, unlike chloroplast, amyloplasts may contain sucrose.[34,35] The similarity of amyloplasts and chloroplasts is not surprising considering that both plastid types are derived from proplastids and often the two plastid types are interconvertible. By using fluorescent-labeled antibodies prepared with highly purified maize endosperm starch branching enzymes, Fisher and Boyer[184] demonstrated the association of branching enzymes with amyloplasts (Figure 1). What remains to be demonstrated, however, is the extent of enzymatic compartmentation in the storage tissue cell. Are all starch biosynthetic enzymes located solely in the plastid? Do amyloplast membranes contain a phosphate translocator involved in the active transfer of triose-P into the organelle? Is the enzyme system for the conversion of triose-P to glucose found in the amyloplast? Is sucrose synthesis and degradation cytosolic, plastid-bound, or both? The answers to these questions should give new insight into the regulation of starch synthesis in storage tissues.

Starch degradation in leaves is usually assumed to occur in the chloroplast. Indeed, most if not all of the starch in leaves has been found to be chloroplastic,[36] and electron micrographs consistently show starch granules as components of chloroplasts. Yet, compartmentation studies show degradative enzymes to be distributed differently.

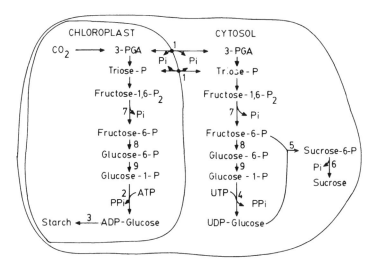

FIGURE 2. Proposed pathway of starch and sucrose biosynthesis from photosynthetically fixed carbon. Enzyme reaction indicated by number are as follows: (1) phosphate translocator; (2) ADP-glucose pyrophosphorylase; (3) starch synthase; (4) UDP-glucose pyrophosphorylase; (5) sucrose-6-phosphate synthase; (6) sucrose-phosphate phosphatase; (7) fructose-6-diphosphatase; (8) glucose phosphate isomerase; and (9) phosphoglucomutase.

Amylases, debranching enzyme, phosphorylase and D-enzyme from spinach leaves were found in both the cytosol and chloroplasts.[26] The major activities, however, were found in the cytoplasm. The major portion of phosphorylase in spinach and pea leaves[37] and all the β-amylase activity of *Vicia faba*[38] and pea shoots[39] were found outside the chloroplast. Purification and characterization of leaf phosphorylase have shown that the cytoplasmic and chloroplastic enzymes differ in molecular weight and kinetic properties and are different enzymes.[40,43] The role of cytoplasmic enzymes in starch degradation at present is unknown. Because starch is made in chloroplasts, the natural assumption is that degradation occurs exclusively in the chloroplast as well. In wheat leaves, starch granules have been observed to be released from chloroplasts to the cytosol through membrane interruptions, particularly during the period of maximum grain filling.[44] Perhaps the release of starch granules from chloroplasts is a general phenomenon at times of accelerated translocation, but no regulation of cytoplasmic enzymes occurs at the level of enzyme synthesis or turnover. The integrity of amyloplast membranes in storage tissues is probably lost during senescence, drying, and storage. Loss of membrane integrity may also occur during the synthetic period of tissue development. However, in seeds in which the storage tissue is nonsenescent, i.e., cotyledons, plastid integrity is maintained as well as function during germination.[10]

B. Starch Synthesis in Leaves

Photosynthetically fixed carbon (3-PGA) can be further metabolized in two different ways (Figure 2). Within the chloroplast, triose-P is converted to hexoses by the enzymes of gluconeogenesis. ADP-glucose is formed from glucose-1-P by the pyrophosphorylysis reaction catalyzed by ADP-glucose pyrophosphorylase (Reaction 2, Figure 2). The glucose moiety of ADP-glucose is then transferred to starch by the starch synthase reaction (Reaction 3, Figure 2). Leaf starch synthases exist as multiple enzyme forms, and are divided by whether they are bound to the starch granule or are soluble. Using DEAE-cellulose chromatography, soluble starch synthases from spinach[45] and maize[46] leaves were also separated into different forms, but only a single form was

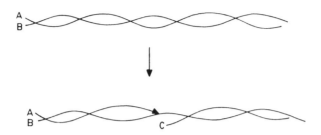

FIGURE 3. Mechanism of starch branching enzyme action. Double helical association of two α-1,4 glucan chains A and B. Left hand ends are nonreducing ends, Chain A is the donor chain, and B is the acceptor chain. Cleavage of an α-1,4 bond in A is made with the subsequent formation of an α-1-6 bond (arrow) and new nonreducing end, C. Chains A and B may be different molecules (interchain transfer) or parts of the same molecule (intrachain transfer).

found in grape leaves.[46] Starch branching enzyme (Q-enzyme) introduces α-1,6 branch points into growing starch molecules by the hydrolysis of an α-1,4 linkage yielding an oligosaccharide (6 to 7 glucose units). The oligosaccharide is then attached to another α-1,4 glucose chain by the formation of an α-1,6 branch point (Figure 3). The transfer of the oligosaccharide has been demonstrated to occur between α-glucans, but intramolecular transfer cannot be ruled out.[47] Branching enzymes from leaves also exist in multiple molecular forms.[46,48]

Photosynthetically fixed carbon (trioses) can be exchanged with inorganic phosphate into the cytoplasm. The exchange is mediated by the phosphate translocator. Hexoses are again synthesized by gluconeogenesis (Figure 2). UDP-glucose pyrophosphorylase produces UDP-glucose which donates the glucose to sucrose-6-P in a reaction catalyzed by sucrose-P synthase (Reaction 5). Finally, sucrose-6-P is hydrolyzed to sucrose by sucrose-6-phosphate phosphatase.

Three reactions are significant in the regulation of the partitioning of photosynthetically fixed carbon in leaves. Both ADP-glucose pyrophosphorylase and sucrose-P synthase are allosteric.[15] In addition, activities of both enzymes are inhibited by inorganic phosphate. As the exchange of phosphate between the cytoplasm and the chloroplast is an integral part of the exchange of carbon (Reaction 1, Figure 3), the phosphate translocator is important in regulating the compartment levels and distribution of Pi.[48]

The kinetic properties of ADP-glucose pyrophosphorylases from a number of bacterial and plant sources have been determined, primarily in the laboratory of Preiss.[15,50-55] The activities of all enzymes were inhibited by Pi, and this inhibition was modulated by 3-PGA. Concentrations of Pi required for 50% inhibition of ADP-glucose pyrophosphorylase activity ranged from 20 to 190 μM, while concentrations of 3-PGA required for 50% activation ranged from 7 to 370 μM. Furthermore, Pi inhibition was observed to be largely overcome by increasing concentrations of 3-PGA. Thus, ADP-glucose pyrophosphorylase activity was sensitive to variations in the concentrations of Pi and 3-PGA as well as the ratio of the two. Because the concentrations of Pi in chloroplasts increases in the dark[56-59] and levels of 3-PGA would be expected to fall with the cessation of photosynthesis, the variation of these effector molecules ([3-PGA]/[Pi]) with physiological condition have been postulated to regulate starch synthesis.

Consistent with this hypothesis are observations of effects of Pi, 3-PGA, or dihydroxyacetone-phosphate (DHAP) on the rates of starch accumulation in leaf discs or isolated chloroplasts. Leaf discs from different plants accumulate higher levels of starch when placed on Pi-deficient media than when incubated on media with normal phosphate levels.[59] Similarly, leaf discs incubated in the light in the presence of mannose also accumulate ten times the level starch as leaf discs incubated in the absence of mannose.[60] Mannose was not incorporated into the starch granule, and thus did not serve as a carbon source. Rather, mannose acted as a phosphate sink through the formation of mannose-6-P which is not further metabolized. Thus, leaf discs incubated in the presence of mannose are also Pi deficient.

The rates of starch synthesis[61-64] and ADP-glucose formation[65,66] in isolated chloroplasts responded to exogenous levels of Pi, 3-PGA, or DHAP the same as leaf discs. Increasing the concentration of Pi from 0.1 mM in the media caused the chloroplast 3-PGA/Pi ratio to fall from 4 to 0.3. A concomitant 25-fold drop in the rate of CO_2 incorporation into starch was observed.[61] In addition, the inhibition of starch synthesis observed in media containing 0.5 mM Pi were reversed by the addition of 3-PGA to the incubation media. The addition of DHAP to the incubation media was similarly found to increase starch synthesis.[64] DHAP can be exchanged by the phosphate translocator as well as 3-PGA, and 3-PGA and DHAP are interconvertible. No inhibition of carbon fixation with the varying ratios or concentrations of Pi, PGA, and DHAP used was observed in this study; thus, it was concluded that starch synthesis could be regulated independently of photosynthesis.

Study of the kinetic properties of sucrose-6-P synthase indicates a possible regulation function. Sucrose has been found to be an inhibitor of wheat germ sucrose-P synthase,[67,68] while the enzyme from spinach leaves was inhibited by sucrose-6-P but not sucrose.[69] Inorganic phosphate inhibited the spinach leaf enzyme with a Ki of 1.75 mM. Therefore, sucrose-P synthase does not appear to be as sensitive to Pi concentration as the ADP-glucose pyrophosphorylases. Although the diurnal variation in enzyme activity is regulated by Pi concentrations, the relative partitioning of photosynthetic carbon between starch and sucrose has been shown to be highly correlated with the level of sucrose-P synthase activity.[70] Species with high sucrose-P synthase activity in the leaves accumulated more sucrose and less starch than species with low activity.

C. Starch Synthesis in Reserve Tissues

Translocated sucrose arriving at the storage organ may be hydrolyzed to glucose and fructose during the unloading and movement into the storage tissue.[71,73] However, the resynthesis of sucrose apparently occurs in reserve tissues as high sucrose levels are observed. The resynthesis of sucrose most likely is catalyzed by sucrose synthase (Reaction 2, Figure 4). However, this enzyme and invertase both participate in sucrose cleavage as well. Based on the metabolite compartmentation studies described above (Section II. A), Shannon[35] presented a general scheme for carbohydrate flow in a cell of reserve tissue and a similar scheme is presented in Figure 4. Triose-P is produced in the cytoplasm and moves into the amyloplast in exchange for Pi via the phosphate translocator (Reaction 10). Hexoses are formed through gluconeogenesis and subsequently used for starch synthesis (or sucrose metabolism?).

Starch synthases from reserve tissue also occur as starch granule-bound and soluble fractions. The starch granule-bound enzyme can use UDP-glucose as well as ADP-glucose as a substrate, though ADP-glucose is the preferred substrate.[50,74,75] Soluble starch synthase has a specific requirement for ADP-glucose and a lower Km for ADP-glucose than the starch granule-bound enzyme. Soluble starch synthases from maize endosperm,[76] potato tubers,[77] rice endosperm,[78] and pea cotyledons[79] have been shown to be present in multiple forms. Similarly, multiple forms of starch branching enzymes have been observed in maize endosperm[80] and developing pea cotyledons,[79] but only a single form was observed in potato tubers.[81] The possibility that the phosphorylase reaction (Reaction 14, Figure 4) may be active in starch synthesis in reserve tissues can

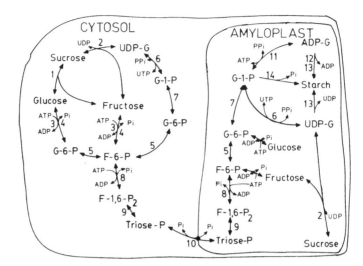

FIGURE 4.   Proposed pathway of starch biosynthesis from translocated sugars in cells of storage organs. Enzyme reactions indicated by number are as follows: (1) invertase; (2) sucrose synthase; (3) hexokinase; (4) hexose-6 phosphatase; (5) glucose phosphate isomerase; (6) UDP-glucose pyrophosphorylase; (7) phosphoglucomutase; (8) phosphofructokinase; (9) aldolase; (10) phosphate translocator; (11) ADP-glucose pyrophosphorylase; (12) soluble starch synthase; (13) starch granule bound starch synthase; and (14) phosphorylase.

not entirely be ruled out at this time. Certainly, phosphorylases are present in developing storage tissue. However, the low level of glucose-1-P in plants and in isolated amyloplasts,[34] would suggest that synthesis of starch through this reaction is highly unlikely (see discussion in References 14 and 34).

The ADP-glucose pyrophosphorylases from reserve tissues are also allosteric.[52,82-88] The enzymes from maize endosperm and potato tubers are the best characterized. The potato enzyme is essentially dependent upon 3-PGA for activity, requiring 0.4 mM 3-PGA for 50% of maximal activity.[88] Similarly, the ratio of 3-PGA and Pi modulate the in vitro activity of this enzyme. However, the ADP-glucose pyrophosphorylase from maize endosperm is less sensitive to Pi and 3-PGA concentrations, with 3 mM Pi required to cause 50% inhibition in the absence of 3-PGA[83] as compared to a range of 0.020 to 0.190 mM Pi found for leaf enzymes. Similarly, the maize enzyme was only slightly stimulated by 3-PGA with 50% stimulation at 2.2 mM. Therefore, the modulation of starch synthesis in reserve tissue by levels of intermediates may be less effectively regulated by metabolite concentrations. Starch synthesis in reserve tissue is also regulated at the level of the synthesis of the starch biosynthetic enzymes. Strong correlation has been shown between starch accumulation and the appearance of ADP-glucose pyrophosphorylase, starch synthase and/or branching enzymes in barley endosperm,[89] maize endosperm,[90,91] potato tubers,[86] potato stolons,[92] pea cotyledons,[79,93,94] and rice grains.[87,95]

D. Pathways of Starch Degradation

Because starch is found in insoluble granules, the initial phase of starch degradation is the enzymatic reduction of the starch granule. Although a number of enzymes are able to attack α-1,4 bonds and debranching enzymes (R-enzymes) can hydrolyze α-1,6 branch points in soluble α-glucans, α-amylase is the only plant enzyme reported to degrade isolated starch granules.[96-102] Starch granules from reserve tissues are more

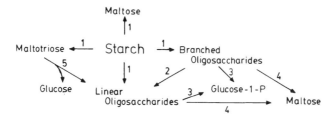

FIGURE 5. Proposed scheme for the degradation of starch. Enzyme reactions indicated by numbers are as follows: (1) α-amylase; (2) debranching enzyme (R-enzyme); (3) phosphorylase; (4) β-amylase; and (5) D-enzyme.

resistant to hydrolysis than leaf starch granules. The reduction of starch granules by α-amylase results in a mixture of maltose, maltotriose, and branched and linear oligosaccharides (Reaction 1, Figure 5). In the second phase of starch degradation, oligosaccharides and maltodextrins are further degradated to maltose, glucose, and glucose-1-P (Figure 5). Finally, maltose, glucose, and glucose-1-P are further metabolized and/or exported. The characteristics of the enzymes of starch degradation have been thoroughly reviewed recently and the reader is referred to these references for further details.[13-15,101-104] Since most of the enzymes of starch degradation are found as multiple isozymic forms, the assignment of specific functions to different isozymic fractions and genetic analysis have been difficult.

The regulation of starch degradation in reserve tissues occurs primarily at the level of enzyme synthesis and activation.[102-104] Enzyme activities increase during seed germination. This increase can occur by two distinct mechanisms, the activation of "latent" enzyme associated with protein bodies to active enzyme, i.e., β-amylase in barley,[105,106] or by *de novo* synthesis, i.e., α-amylases in cereal aleurones.[107-111] In addition to the well-characterized α-amylase response of cereal aleurone to gibberellins, α-amylase (and β-amylase) are also synthesized and secreted from scutellum epithelial cells.[112-113] This secretion in rice seeds has been demonstrated to occur prior to the aleurone production of α-amylase. The α-amylase in legume seeds[114-117] are also synthesized *de novo*, while β-amylase and debranching activities appear to be released from latent forms associated with protein particles.[117-120]

Starch degradation in leaves appears to occur through both phosphorylytic and amylolytic activities.[60-63,121-124] These activities are well documented in leaves and analysis of degradation products in a number of studies and under varying conditions suggest the presence of both pathways.[60-63,121-124] Regulation of starch degradation in leaves is not clearly understood. The coordinate synthesis and degradation of starch may in part be regulated by differing pH and Pi concentrations of chloroplasts during light and dark periods. For example, high Pi concentrations would inhibit ADP-glucose formation but would be necessary for phosphorylase activity.

III. GENETIC MODIFICATION OF STARCH METABOLISM

Present evidence demonstrates that starch metabolism can be varied by mutation. These mutations are usually recessive and of spontaneous origin. The recessive nature of the mutations makes the detection of homologous mutants in polyploid species unlikely. In this section, known mutations in the starch biosynthetic pathway will be discussed in relation to the lesion in the pathway. Large reductions in starch quantity are generally associated with mutations of enzymes early in the pathway. Smaller reductions in starch, but significantly altered ratios or structures of amylopectin and amylose are characteristic of starch synthase and starch branching enzyme mutants.

Table 1
PROPERTIES OF MAIZE KERNELS HOMOZYGOUS FOR
MAIZE ENDOSPERM GENES

| Genotype | Kernel phenotype[a] | Enzyme deficiency | Starch (% of dry wt) | |
			Immature endosperm	Mature endosperm
Nonmutant	Full — translucent	None	80	83
Brittle	Shrunken, opaque to tarnished	Unknown	17	31—47
Brittle-2	Shrunken, opaque to tarnished	ADP-glucose pyrophosphorylase	7	28—46
Shrunken	Collapsed, opaque	Sucrose synthase	60	77
Shrunken-2	Shrunken, opaque to translucent	ADP-glucose pyrophosphorylase	16—21	54
Shrunken-4	Shrunken, opaque	Reduced pyridoxal phosphate synthesis	52	66—74

[a]   Based on descriptions by Garwood and Creech.[125]
[b]   Data compiled from a number of sources[16,126-128] and represent reported ranges.

Presently, few mutations are known for starch degradative enzymes and possible reasons for this will be discussed. Finally, the use of starch mutants in plant breeding will be described.

A. Mutations Early in the Starch Biosynthetic Pathway

The developing endosperm of maize has been used in genetic studies for nearly 80 years. During this period, a large number of genes altering starch synthesis have been described. Mutations of enzymes in the early part of the starch biosynthetic pathway have been observed and the properties of the kernels of these genotypes are compared in Table 1. Two mutations, *shrunken-2* and *brittle-2,* reduce the ADP-glucose pyrophosphorylase activity to 5 to 10% of the level found in nonmutant.[128-131] The *shrunken (sh)* gene has been demonstrated to be the structural gene for sucrose synthase.[131] A low level of sucrose synthase activity (10% of nonmutant endosperm) was detected in homozygous *sh* endosperm. Based on kinetic properties, this activity is thought to be a minor sucrose synthase enzyme.[133] The significance of minor isozymes is unknown at this time. However, minor isozymes have also been observed for maize endosperm starch granule-bound starch synthase[134] and pea cotyledon branching enzyme.[94] Therefore, the presence of minor isozymes may be a general phenomenon. The *shrunken-4 (sh-4)* mutant was originally thought to be a phosphorylase mutant.[127] A latter investigation, however, revealed that the *sh-4* lesion was in the pathway for the formation of pyridoxal phosphate.[135] Since pyridoxal phosphate is a cofactor for many enzymes, including phosphorylase, a number of endosperm enzyme activities would be expected to be lower in vivo. A final maize mutant which severely reduces endosperm starch content is *brittle (bt).* To date, no enzymatic explanation for *bt* is known.

Presently, mutants homologous to the above maize mutants are only known at the level of starch reductions and altered phenotypes. Shrunken mutants of barley with maternal control of seed character have been reported.[136] Similarly, the *rb* mutant of peas reduces the level of starch significantly, but produces cotyledon starch with a normal ratio of amylose and amylopectin.[137] A more thorough and conscientious effort to characterize presently available or induced mutants of shrunken types in additional species should result in the identification of mutants homologous to the maize loci. It will be important to the understanding of starch biosynthesis and its genetic control that the characterization of these mutants includes studies at the enzymatic level.

Table 2
PROPERTIES OF STARCH SYNTHASE AND BRANCHING ENZYME MUTANTS OF MAIZE ENDOSPERM

Genotype	Kernel phenotype[a]	Enzyme deficiency	Amylose[b] (%)	Phytoglycogen
Nonmutant	Full — translucent	None	28.9	–
Waxy	Opaque	Starch granule bound starch synthase	0.4	–
Amylose-extender	Tarnished — translucent	Branching enzyme IIb	59.8	–
Dull	Opaque to tarnished	Branching enzyme IIa: soluble starch synthase II (both decreased)	34.2	–
Sugary-2	Slightly tarnished	Unknown	37.8	–
Sugary	Wrinkled — glassy	Unknown	32.6	+

[a] Based on the descriptions of Garwood and Creech.[125]
[b] Adapted from Holder et al.[157] and are averages of ears from six different years.

B. Starch Synthase and Starch Branching Enzyme Mutations

Mutations with altered amylose:amylopectin ratios or structural differences in the α-glucans fall into three classes. In the first class of mutants (usually termed *waxy* or *glutinous*), starch granules contain essentially 100% amylopectin. The second class of mutants produces unusually high amounts of amylose, ranging from 40 to 70% of the starch, and are termed high amylose mutants. Finally, *sugary* mutants produce higher levels of sucrose but reduced levels of starch. In addition to starch, *sugary* mutants contain significant quantities of the water-soluble polysaccharide phytoglycogen which is similar in structure to glycogen. Examples of all three classes of mutants are known for maize endosperm and the properties of these mutants are summarized in Table 2.

Waxy (wx) or *glutinous (gl)* mutations are known in a wide range of grass species including maize,[138] rice,[139] barley,[140] sorghum,[141] Job's tears (*Coix lachryma-jobi*),[142] foxtail millet (*Setaria italica*),[143] and proso millet (*Panicum miliaceum*).[143] In addition, *wx* mutations have been observed in the perisperm of amaranth *(Amaranthus* spp.) seeds.[143-147] Because *wx* starch granules stain reddish brown with iodine compared to the blue-black color of starch granules containing amylose, it is not surprising that these mutants have been detected frequently. Murata et al.[148] reported that *wx* rice lacked starch granule-bound starch synthase. Starch granule-bound starch synthase is also lacking in *wx* maize.[149] *Waxy* mutants show clear dosage effects for amylose content and starch granule-bound starch synthase activity [147,150-152] and thus have been concluded to be structural genes for the starch granule-bound synthase. This conclusion is supported by the recent observation that the *waxy* protein was the predominant protein in extracts from nonmutant starch granules, but absent in extracts from starch granules from *waxy* kernels.[153]

High amylose mutants have been reported for maize endosperm,[154] pea cotyledons,[155] and barley endosperm.[156] The *amylose-extender* gene of maize, the *r* gene of peas and *amy-1* gene in barley all condition increases in the amylose content from 25 to 30% in the nonmutant to 50% or higher. Amylose and amylopectin from *ae, r,* and *amy-1* starches are quite different from nonmutant amylose and amylopectin. The amylose fractions have a lower average molecular weight and larger range of sizes than nonmutant amylose (see References 2 and 22 for review). Furthermore, the amylopectin fractions differ from normal amylopectins by being less highly branched.[160-164] The *ae* and *r* loci have been demonstrated to be branching enzyme mutants.[94,165] In the case of *ae,* based on dosage studies, we suggest that *ae* is a structural gene for branching enzyme IIb.[166-168] The failure to produce a completely linear α-glucan in these mutants is due to the fact that *ae* endosperms contain two independent branching enzymes and *r* cotyledons also contain a minor branching enzyme which are present and active. In maize, two other genes, *dull* and *sugary-2,* also cause modest increases in the amylose content of endosperm starch (Table 2).[157] Endosperm homozygous for *dull* contain reduced levels of both soluble starch synthase and starch branching enzyme activities.[158,159] This observation is not easily interpreted but suggests a possible enzymatic explanation for the effect of *dull.* Although the amylose content is higher in *dull* starch, the structure of the amylose and amylopectin are the same as in nonmutant starch.

Phytoglycogen is a well-known component of *sugary* maize kernels. Phytoglycogen is highly branched and is distributed between particulate and soluble fractions.[9,170] Karper and Quimby[171] reported a homologous mutant in sorghum which was later found to contain high levels of water-soluble polysaccharides.[172] Unpublished results from my laboratory have now demonstrated that this water-soluble polysaccharide is structurally similar to *sugary* maize phytoglycogen. The phytoglycogen from *sugary* sorghum was found in both particulate and soluble fractions and an average chain-length of 12.0 glucose units was found for sorghum phytoglycogen. Several studies have suggested a unique branching enzyme as the explanation for phytoglycogen in *sugary* maize.[173-175] However, we recently reported that nonmutant kernels also contain a branching enzyme (branching enzyme I) capable for forming phytoglycogen in in vitro reactions.[176] Therefore, a more appropriate question to be answered at present is not "Why is phytoglycogen formed in *sugary* kernels?" but "Why is phytoglycogen not formed in nonmutant kernels?"

In all mutants examined to date, clear tissue specificity has been observed. Generally speaking, early pathway mutations are expressed only in the endosperm of maize seed.[177] Starch synthase and branching enzyme mutants are expressed in the endosperm (or cotyledon) within the seed, and are also usually expressed in pollen as well.[139,178-179] No evidence has been reported to indicate the expression of these mutations in the leaves.

C. Mutations of Starch Degradation

Clear cut examples of mutations of starch degradation are rare. Adams et al.[10] reported that two soybean cultivars "Chestnut" and "Altona" contained very low levels of β-amylase. Yet, no obvious effect on starch metabolism could be observed. As pointed out above (Section II.D), starch degradative enzymes are usually present in multiple forms. Although isozymic variability exists for phosphorylases and amylases, no sequential pattern of isoenzyme action has been described. Therefore, no clear phenotype differences may arise in mutations of starch degradative enzymes due to the multiple activity, thus making these mutants less obvious than synthetic mutations. Similarly, mutations in starch degradation may not be obvious until germination and lethality is one possible consequence, which, if expressed during germination, would not be observed in segregating population. Similarly, selection for early germination would be expected to eliminate many mutations.

D. Present and Potential Use of Starch Mutants in Plant Breeding

The value of some mutants affecting starch accumulation was realized before the practice of modern plant breeding. The use of *sugary* corn by native Americans is well documented.[18,180] As this mutation is recessive, the value of *sugary* corn was sufficiently great for the practice of isolation which is necessary to maintain the genotype to be developed. Presently the *sugary* gene forms the basis of standard sweet corn cultivars. Today sweet corn cultivars with improved sweetness and retention of sweetness both on the plant and postharvest have been and are being developed using a number of the above endosperm mutants both individually and in combination.[18] Similarly, the *waxy* mutant is the major gene determining the preferred "glutinous" character of many rice cultivars. Based on the increasing evidence for homologous mutants from divergent species, I would suggest that similar improvements in the "quality" of other crops or the development of "specialty crops" can be achieved. Although many crops (i.e., sorghum and amaranths) are primarily used for livestock feed in the U. S., these crops are used both green and mature in other cultures for human consumption. Therefore, sweet cultivars of these crops may be developed in the future. Equally important in developing nations, however, is the efficiency of calorie use. Starch granules from *waxy* endosperms have been repeatedly observed to be more readily digested that nonmutant granules in in vitro assays. As *waxy* cultivars have similar yield potential as nonmutant cultivars, the use of *waxy* mutants in feed grain cultivars is being promoted by some U.S. seed companies. The potential of these mutants for human nutrition needs a closer examination.

The *amylose-extender* and *waxy* mutants form the basis of hybrids for industrial purposes.[19] Hybrids homozygous for *amylose-extender* are currently produced. Despite lower yield, these hybrids are grown as a source of high amylose starch under contract as a specialty corn. *Waxy* hybrids are also produced as a source of starch which is used as a component for the food industry (baby food, frozen food, etc.). The potential of novel starches resulting from multiple gene combinations continues to be explored in several laboratories.[17]

Further development and use of cultivars with starch mutants is needed in the future. Continued interest in improved quality of foods will necessitate the further exploration of existing and new genotypes. The decreased availability and increased cost of petroleum feed stocks for the chemical industry make starch an attractive, renewable alternative,[4] but this possibility needs further evaluation. The problems presented to the plant breeder are not great. As these mutants are relatively easy to distinguish based on phenotype and inherited as recessive genes, their use in backcross breeding programs or other breeding procedures is not complicated. The major obstacle encountered will be the need to fully evaluate the germplasm available for development of cultivars. Expression of the mutants in terms of reduction in germination, seedling vigor, and yield is widely variable. Therefore, development of suitable hybrids will depend on the evaluation of a wide range of germplasm. Regardless of the planned use of a cultivar, the value of the product will need to offset any reduction in yield encountered. In order to fully accomplish these objectives, the cooperation and participation of biochemists, geneticists, and plant breeders will be necessary.

IV. QUANTITATIVE VARIATION IN STARCH METABOLISM

The variation in the amount of starch found in different cultivars is not surprising. The accumulation of starch in seeds (or leaves) will vary with different rates of photosynthesis, phloem loading, translocation, unloading of translocated sugars at the sink, and finally the rate of starch synthesis and sink size. Shannon[35] has recently described the complexity of potential processes which may be rate limiting in crop productivity. In addition, the levels of starch in cereal endosperm fluctuate with varying protein levels keeping the proportion of protein and starch relatively constant.[16] With this complexity in mind, it would seem risky to conclude that differences in starch levels are only due to quantitative variation in starch metabolism. Increased understanding

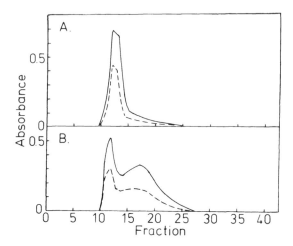

FIGURE 6.   Gel filtration elution profiles of starches isolated from maize kernels homozygous for the endosperm mutants *amylose-extender* and *waxy*. (A) W64A inbred; (B) Ia5125 inbred. A_{540}—; A_{660}---.

of the physiology of crop productivity is needed before the significance of the variation of starch levels can be fully evaluated.

We have demonstrated quantitative variation in starch metabolism through a different approach. The effects of mutations of starch synthesis and branching enzymes in maize endosperm have major effects on the properties of the starch. For example, *waxy* endosperms contain amylose free starch. This condition is observed when other mutants are also incorporated into double or triple mutant combinations. However, the molecular weight distribution of the starches from the same genotype in different backgrounds vary. The size distribution variation can range from unimodal to biomodal distributions for the same genotype (Figure 6).[163,181] The effect of the major gene is not altered, but additional variation exists. Similar variation is also observed for starch granule (plastid) morphology for these gene combinations.[182,183] This observation implies that minor gene differences can change starch granule structure as well as the structure of amylose and amylopectin. Presently, the significance of this variation is unclear. In addition, no simple screening techniques have been developed to enable plant breeders to utilize this variation.

V. SUMMARY

Through the research presented in this review, I have attempted to describe the metabolism of starch in higher plants in a unified manner. Although, differences exist between enzymes of starch metabolism between tissues of the same species or between different species, biochemical research on compartmentation and enzymology supports the concept of conserved pathways for starch metabolism. The study of genetic mutants of starch metabolism has had a synergistic effect, increasing the understanding of both the biochemistry and the genetic control of starch metabolism. It is significant that the growing number of mutants of starch metabolism indicate homologies of mutants in different species. Continued characterization of starch mutants from a wide range of species should be a rewarding endeavor in the future.

Although research on starch metabolism and its genetic control remain active areas of study, it should be apparent to the reader that deficiencies in our basic knowledge continue to exist. Although these deficiencies exist, the potential, both realized and future, of starch mutants as the raw material for plant breeding should continue to expand. The prospect for future gains in our basic knowledge and its application will be benefited by coordinated concentrated investigations. Cooperative research efforts will require the combined expertise of biochemists, physiologists, geneticists, and plant breeders.

REFERENCES

1. Badenhuizen, N. P., Occurrence and development of starch in plants, in *Starch: Chemistry and Technology,* Vol. 1, Whistler, R. L. and Paschall, E. F., Eds., Academic Press, New York, 1965, 65.
2. Banks, W. and Muir, D. D., Structure and chemistry of the starch granule, in *The Biochemistry of Plants,* Vol. 3, Preiss, J., Ed., Academic Press, New York, 1980, chap. 9.
3. Hizukuri, S., Takdea, Y., Yasuda, M., and Suzuki, A., Multi-branched nature of amylose and the action of debranching enzymes, *Carbohydr. Res.,* 94, 205, 1981.
4. Von Woelk, H. V., Starke als chemicrohstoff moglichkeiten und grenzen, *Starke,* 33, 397, 1981.
5. Craig, J. S., Storage products, in *Algal Physiology and Biochemistry,* Stewart, W. D. P., Ed., Blackwell Scientific Publ., Oxford, 1974, 206.
6. Frederick, J. S., Glucosyltransferase isozymes in algae. II. Properties of branching enzymes, *Phytochemistry,* 7, 931, 1968.
7. Meeuse, J. D., Andries, M., and Wood, J. A., Floridean starch, *J. Exp. Bot.,* 11, 129, 1960.
8. Stacey, M. and Barker, S. A., *Polysaccharides of Micro-organisms,* Oxford University Press, Oxford, 1960.
9. Stitt, M. and Heldt, H. W., Simultaneous synthesis and degradation of starch in spinach chloroplasts in the light., *Biochim, Biophys. Acta,* 638, 1, 1981.
10. Adams, C. A., Rinne, R. W., and Fjerstad, M. C., Starch deposition and carbohydrase activities in developing and germinating soya bean seeds, *Ann. Bot.,* 45, 577, 1980.
11. Evans, L. T., The natural history of crop yield, *Am. Sci.,* 68, 388, 1980.
12. Hegsted, D. M., Protein-calorie malnutrition, *Am. Sci.,* 66, 61, 1978.
13. Preiss, J. and Levi, C., Metabolism of starch in leaves, in *Encyclopedia of Plant Physiology,* Vol. 6, Gibbs, M. and Latzko, E., Eds., Springer-Verlag, Berlin, 1979, 282.
14. Preiss, J. and Levi, C., Starch biosynthesis and degradation, in *The Biochemistry of Plants,* Vol. 3, Preiss, J., Ed., Academic Press, New York, 1980, 371.
15. Preiss, J., Regulation of the biosynthesis and degradation of starch, in *Annu. Rev. Plant Physiol.,* 33, 431, 1982.
16. Nelson, O. E., Genetic control of polysaccharide and storage protein synthesis in the endosperms of barley, maize and sorghum, in *Adv. Cereal Sci. Tech.,* 3, 41, 1981.
17. Shannon, J. C. and Garwood, D. L., Genetics and physiology of starch development, in *Starch: Chemistry and Industry,* Whistler, R. L., Paschall, E. F., and BeMiller, J. N., Eds., Academic Press, New York, 1984, 26.
18. Boyer, C. and Shannon, J. C., The use of endosperm mutants for sweet corn improvements, *Plant Breeding Rev.,* 1, 139, 1982.
19. Alexander, D. E. and Creech, R. G., Breeding special industrial and nutritional types, in *Corn and Corn Improvement,* Sprague, G. F., Ed., American Society of Agronomy, Madison, Wisc., 1977, chap. 7.
20. Badenhuizen, N. P., *The Biogenesis of Starch Granules in Higher Plants,* Appleton-Century-Crofts, New York, 1969.
21. Radley, J. A., *Starch and its Derivatives,* 4th ed., Chapman and Hall, Ltd., London, 1968.
22. Banks, W. and Greenwood, C. T., *Starch and Its Components,* Edinburgh University Press, Edinburgh, 1975.
23. Leloir, L. F., de Fekete, M. A. R., and Cardini, C. E., Starch and oligosaccharide synthesis from uridine diphosphate glucose, *J. Biol. Chem.,* 236, 636, 1961.
24. Jensen, R. G., Biochemistry of the chloroplast, in *The Biochemistry of Plants,* Vol. 1, Tolbert, N. R., Ed., Academic Press, New York, 1980, 274.
25. Mares, D. J., Hawker, J. S., and Possingham, J. V., Starch synthesizing enzymes in chloroplasts of developing leaves of spinach *(Spinacea oleracea* L.), *J. Exp. Bot.,* 29, 829, 1978.

26. Okita, T. W., Greenberg, E., Kuhn, D. N., and Preiss, J., Subcellular localization of the starch degradative and biosynthetic enzymes of spinach leaves, *Plant Physiol.*, 64, 187, 1979.
27. Bird, I. F., Cornelius, M. J., Keys, J., and Whittington, C. P., Intracellular site of sucrose synthesis in leaves, *Phytochemistry*, 13, 59, 1974.
28. Williams, B. R., The Ultrastructure of Plastid Differentiation in Endosperm Cells of *Normal* and *Sugary-1 Zea mays* L., Ph. D. thesis, The Pennsylvania State University, University Park, 1971.
29. Buttrose, M. S., Submicroscopic development and structure of starch granules in cereal endosperms, *J. Ultrastr. Res.*, 4, 231, 1960.
30. Ohad, I., Friedberg, I., Ne'eman, Z., and Schramm, M., Biogenesis and degradation of starch. I. The fate of the amyloplast membranes during maturation and storage of potato tubers, *Plant Physiol.*, 47, 465, 1971.
31. Newcomb, E., Fine structure of protein-storing plastids in bean root tips, *J. Cell Biol.*, 33, 143, 1967.
32. Whistler, R. L. and Thomburg, U. L., Starch granule formation: development of starch granules in corn endosperm, *J. Agr. Food Chem.*, 5, 203, 1957.
33. Liu, T.-T. Y. and Shannon, J. C., A nonaqueous procedure for isolating starch granules with associated metabolites from maize (*Zea mays* L.) endosperm, *Plant Physiol.*, 67, 518, 1981.
34. Liu, T.-T. Y. and Shannon, J. C., Measurement of metabolites associated with nonaqueously isolated starch granules from immature *Zea mays* L. endosperm, *Plant Physiol.*, 67, 525, 1981.
35. Shannon, J. C., A search for rate-limiting enzymes that control crop production, *Iowa State J. Res.*, 56, 307, 1982.
36. Stitt, M., Bulpin, P. V., and ApRees, T., Pathway of starch breakdown in photosynthetic tissues of *Pisum sativum, Biochim. Biophys. Acta*, 544, 200, 1978.
37. Steup, M. and Latzko, E., Intercellular localization of phosphorylases in spinach and pea leaves, *Planta*, 145, 69, 1979.
38. Chapman, G. W., Pallas, J. E., and Mendocini, J., The hydrolysis of maltodextrins by a α-amylase isolated from leaves of *Vicia faba, Biochim. Biophys. Acta*, 276, 491, 1972.
39. Levi, C. and Preiss, J., Amylopectin degradation in pea chloroplast extracts, *Plant Physiol.*, 61, 218, 1978.
40. Preiss, J., Okita, T. W., and Greenberg, E., Characterization of spinach leaf phosphorylases, *Plant Physiol.*, 66, 864, 1980.
41. Steup, M., Schachtele, C., and Latzko, E., Purification of a non-chloroplastic α-glucan phosphorylase from spinach leaves, *Planta*, 148, 168, 1980.
42. Steup, M., Schachtele, C., and Latzko, E., Separation and partial characterization of chloroplast and non-chloroplast α-glucan phosphorylase from spinach leaves, *Z. Pflanzenphysiol.*, 96, 365, 1980.
43. Steup, M., Purification of chloroplast α-1,4-glucan phosphorylase from spinach leaves by chromatography on Sepharose-bound starch, *Biochim. Biophys. Acta*, 659, 123, 1981.
44. Repka, J., Marek, J., Hraska, S., and Bezo, M., Starch release from chloroplasts of wheat flag leaves, *Photosynthetica*, 15, 256, 1981.
45. Ozbun, J. L., Hawker, J. S., and Preiss, J., Soluble adenosine diphosphate glucose-α-1,4-glucan α-4-glucosyltransferases from spinach leaves, *Biochem. J.*, 126, 953, 1972.
46. Hawker, J. S. and Downton, W. J. S., Starch synthases from *Vitis vinifera* and *Zea mays*, *Phytochemistry*, 13, 893, 1974.
47. Borovsky, D., Smith, E. E., and Whelan, W. J., On the mechanism of amylose branching by potato Q-enzyme, *Eur. J. Biochem.*, 62, 307, 1976.
48. Hawker, J. S., Ozbun, J. L., Ozaki, H., Greenberg, E., and Preiss, J., Interaction of spinach leaf adenosine diphosphate glucose α-1,4-glucan α-4-glucosyl transferase and α-1,4-glucan, α-1,4-glucan-6-glycosyl transferase in synthesis of branched α-glucan, *Arch. Biochem. Biophys.*, 160, 530, 1974.
49. Heber, V. W. and Heldt, H. W., The chloroplast envelop: structure, function and role in leaf metabolism, *Annu. Rev. Plant Physiol.*, 32, 139, 1981.
50. Ghosh, H. P. and Preiss, J., The biosynthesis of starch in spinach chloroplasts, *J. Biol. Chem.*, 240, 960, 1965.
51. Ghosh, H. P. and Preiss, J., Adenosine diphosphate glucose pyrophosphorylase: a regulatory enzyme in the biosynthesis of starch in spinach leaf chloroplasts, *J. Biol. Chem.*, 241, 4491, 1966.
52. Preiss, J., Ghosh, H. P., and Wittkop, J., Regulation of the biosynthesis of starch in spinach leaf chloroplasts, in *Biochemistry of Chloroplasts*, Vol. 2, Goodwin, T. W., Ed., Academic Press, New York, 71, 1967.
53. Sanwal, G. G., Greenberg, E., Hardie, J., Cameron, E. C., and Preiss, J., Regulation of starch biosynthesis in plant leaves: activation and inhibition of ADPglucose pyrophosphorylase, *Plant Physiol.*, 43, 417, 1968.
54. MacDonald, P. W. and Strobel, G. A., Adenosine diphosphate-glucose pyrophosphorylase control of starch accumulation in rust-infected wheat leaves, *Plant Physiol.*, 46, 126, 1970.
55. Hawker, J. S., Effect of temperature on lipid, starch and enzymes of starch metabolism in grape, tomato, and broad bean leaves, *Phytochemistry*, 21, 33, 1982.

56. Heber, U. W. and Santarius, K. A., Compartmentation and reduction of pyridine nucleotides in relation to photosynthesis, *Biochim. Biophys. Acta,* 109, 390, 1965.
57. Santarius, K. A. and Heber, U. W., Changes in the intracellular levels of ATP, ADP, AMP, and Pi and regulatory function of the adenylate system in leaf cells during photosynthesis, *Biochim. Biophys. Acta,* 102, 39, 1965.
58. Heber, U. W., Transport metabolites in photosynthesis, in *Biochemistry of Chloroplasts,* Vol. 2, Goodwin, T. W., Ed., Academic Press, New York, 1967, 71.
59. Chen-She, S. H., Lewis, D. H., and Walker, D. A., Stimulation of photosynthetic starch formation by sequestration of cytoplasmic orthophosphate, *New Phytol.,* 74, 383, 1975.
60. Herold, A., Lewis, D. H., and Walker, D. A., Sequestration of cytoplasmic orthophosphate by mannose and its differential effect on photosynthesis in C_3 and C_4 species, *New Phytol.,* 76, 397, 1976.
61. Heldt, H. W., Chen, C. J., Maronde, D., Herold, A., Stankovic, Z. S., Walker, D. A., Kraminer, A., Kirk, M. R., and Heber, U., Role of orthophosphate and other factors in the regulation of starch formation in leaves and isolated chloroplasts, *Plant Physiol.,* 59, 1146, 1977.
62. Steup, M., Peavey, D. G., and Gibbs, M., The regulation of starch metabolism by inorganic phosphate, *Biochem. Biophys. Res. Commun.,* 72, 1554, 1976.
63. Peavey, D. G., Steup, M., and Gibbs, M., Characterization of starch breakdown in isolated spinach chloroplasts, *Plant Physiol.,* 60, 305, 1977.
64. Portis, A. R., Jr., Effects of the relative extrachloroplastic concentrations of inorganic phosphate, 3-phosphoglycerate and dihydroxyacetone phosphate on the rate of starch synthesis in isolated spinach chloroplasts, *Plant Physiol.,* 70, 393, 1982.
65. Kaiser, W. M. and Bassham, J. A., Light-dark regulation of starch metabolism in chloroplasts. I. Levels of metabolites in chloroplasts and medium during light-dark transition, *Plant Physiol.,* 63, 105, 1979.
66. Kaiser, W. M. and Bassham, J. A., Light-dark regulation of starch metabolism in chloroplasts. II. Effects of chloroplastic metabolite levels on the formation of ADP glucose by chloroplast extracts, *Plant Physiol.,* 63, 109, 1979.
67. Salerno, G. L. and Pontis, G. L., Studies on sucrose-phosphate synthetase. Inhibitory action of sucrose, *FEBS Lett.,* 86, 263, 1978.
68. Salerno, G. L. and Pontis, G. L., Regulation of sucrose levels in plant cells, in *Mechanisms of Polysaccharide Polymerization and Depolymerization,* Marshall, J. J., Ed., Academic Press, New York, 1980, 31.
69. Amir, J. and Preiss, J., Kinetic characterization of sucrose-P synthase from spinach leaves, *Plant Physiol.,* 69, 1027, 1982.
70. Huber, S. C., Interspecific variation in activity and regulation of leaf sucrose-P synthetase, *Z. Pflanzenphysiol.,* 102, 443, 1981.
71. Hawker, J. S. and Hatch, M. D., Mechanism of sugar storage by mature stem tissue of sugar cane, *Physiol. Plant,* 18, 444, 1965.
72. Shannon, J. C. and Dougherty, C. T., Movement of ^{14}C-labeled assimilates into kernels of *Zea mays* L. II. Invertase activity of the pedicel and placento-chalazal tissues, *Plant Physiol.,* 49, 203, 1972.
73. Russell, C. R. and Morris, D. A., Invertase activity, soluble carbohydrates and inflorescence development in the tomato (*Lycopersicon esculentum* Mill.), *Ann. Bot.,* 49, 89, 1982.
74. Recondo, E. and Leloir, L. F., Adenosine diphosphate glucose and starch synthesis, *Biochem. Biophys. Res. Commun.,* 6, 85, 1961.
75. Frydman, R. B. and Cardini, C. E., Studies on the biosynthesis of starch. I. Isolation and properties of the soluble adenosine diphosphate glucose:starch glucosyltransferase of *Solanum tuberosum,* *Arch. Biochem. Biophys.,* 116, 9, 1966.
76. Ozbun, J. L., Hawker, J. S., and Preiss, J., Adenosine diphosphoglucose-starch glucosyltransferases from developing kernels of waxy maise, *Plant Physiol.,* 48, 765, 1971.
77. Hawker, J. S., Ozbun, J. L., and Preiss, J., Unprimed starch synthesis by soluble ADP glucose-starch glucosyltransferase from potato tubers, *Phytochemistry,* 11, 1287, 1972.
78. Pisigan, R. A. and Del Rosario, E. J., Isozymes of soluble starch synthases from *Oryza sativa* grains, *Phytochemistry,* 15, 71, 1976.
79. Matters, G. L. and Boyer, C. D., Starch synthases and branching enzymes from *Pisum sativum,* *Phytochemistry,* 20, 1805, 1981.
80. Boyer, C. D. and Preiss, J., Multiple forms of (1-4)-α-D-glucan, (1-4)-α-D-glucan-6-glycosyl transferase from developing *Zea mays* L. kernels, *Carbohydr. Res.,* 61, 321, 1978.
81. Drummond, G. S., Smith, E. E., and Whelan, W. J., Purification and properties of potato α-1,4-glucan α-1,4-glucan 6-glucosyltransferase (Q-enzyme), *Eur. J. Biochem.,* 26, 168, 1972.
82. Amir, J. and Cherry, J. H., Purification and properties of adenosine diphosphoglucose pyrophosphorylase from sweet corn, *Plant Physiol.,* 49, 493, 1972.

83. Dickinson, D. and Preiss, J., ADP-glucose pyrophosphorylase from maize endosperm, *Arch. Biochem. Biophys.*, 130, 119, 1969.

84. Hannah, L. C. and Nelson, O. E., Characterization of adenosine diphosphate glucose pyrophosphorylases from developing maize seeds, *Plant Physiol.*, 55, 297, 1975.

85. Preiss, J., Lammel, C., and Sabrow, A., A unique adenosine diphosphoglucose pyrophosphorylase associated with maize embryo tissue, *Plant Physiol.*, 47, 104, 1971.

86. Sowonkinos, J. R., Pyrophosphorylases in *Solanum tuberosum*. I. Changes in ADP glucose pyrophosphorylase activities associated with starch biosynthesis during tuberization, maturation and storage of potatoes, *Plant Physiol.*, 57, 63, 1976.

87. Sowonkinos, J. R., Pyrophosphorylases in *Solanum tuberosum*. II. Catalytic properties and regulation of ADP glucose and UDP glucose pyrophosphorylase activities in potatoes, *Plant Physiol.*, 68, 924, 1981.

88. Sowonkinos, J. R. and Preiss, J., Pyrophosphorylases in *Solanum tuberosum*. III. Purification, physical and regulatory properties of potato tuber ADP glucose pyrophosphorylase, *Plant Physiol.*, 69, 1459, 1982.

89. Baxter, E. D. and Duffus, C. M., Starch synthetases in developing barley amyloplasts, *Phytochemistry*, 10, 2641, 1971.

90. Ozbun, J. L., Hawker, J. S., Greenberg, E., Lammel, C., Preiss, J., and Lee, E. Y. C., Starch synthase, phosphorylase, ADP glucose pyrophosphorylase and UDP glucose pyrophosphorylase in developing maize kernels, *Plant Physiol.*, 51, 1, 1973.

91. Tsai, C. Y., Salamini, F., and Nelson, O. E., Enzymes of carbohydrate metabolism in the developing endosperm of maize, *Plant Physiol.*, 46, 299, 1970.

92. Obata-Sasamota, H. and Suzuki, H., Activities of enzymes relating starch synthesis and endogenous levels of growth regulators in potato stolon during tuberization, *Physiol. Plant*, 45, 320, 1979.

93. Turner, J. F., Physiology of the pea fruit. VI. Changes in uridine diphosphate glucose pyrophosphorylase and adenosine diphosphate glucose pyrophosphorylase in the developing seed, *Aust. J. Biol. Sci.*, 22, 1145, 1969.

94. Matters, G. L. and Boyer, C. D., Soluble starch synthase and starch branching enzymes from cotyledons of smooth and wrinkle seeded lines of *Pisum sativum* L., *Biochem. Genet.*, 20, 833, 1982.

95. Perez, C. M., Perdon, A. A., Resurrection, A. P., Villareal, R. M., and Juliano, B. O., Enzymes of carbohydrate metabolism in the developing rice grain, *Plant Physiol.*, 56, 579, 1975.

96. Whelan, W. J., Starch and similar polysaccharides, in *Encyclopedia of Plant Physiology*, Vol. 6, Ruhland, Ed., Springer-Verlag, Berlin, 1958, 155.

97. Walker, G. J. and Hope, P. M., The action of some α-amylases on starch granules, *Biochem. J.*, 86, 452, 1963.

98. Leach, H. W. and Schoch, T. J., Structure of the starch granule. II. Action of various amylases on granular structure, *Cereal Chem.*, 38, 34, 1961.

99. Dunn, G., A model for starch breakdown in higher plants, *Phytochemistry*, 13, 1341, 1974.

100. Bailey, R. W. and Macrae, J. C., Hydrolysis of intact leaf starch granules by glycoamylase and α-amylase, *FEBS Lett.*, 31, 203, 1973.

101. Manners, D. J., Some aspects of the enzymic degradation of starch, in *Plant Carbohydrate Biochemistry*, Pridham, J. B., Ed., Academic Press, New York, 1974, 109.

102. Manners, D. J., The structure and metabolism of starch, *Essays Biochem.*, 10, 37, 74.

103. ApRees, T., Pathways of carbohydrate breakdown in higher plants, in *Plant Biochemistry*, Vol. 2, Northcote, D. H., Ed., MTP Int. Rev. Sci. Biochem., Butterworth, London, and University Press, Baltimore, 1974, 89.

104. Briggs, D. E., Hormones and carbohydrate metabolism in germinating cereal grains, in *Biosynthesis and Its Control in Plants*, Milborrow, B. C., Ed., Academic Press, New York, 1973, 219.

105. Hejgaard, J., "Free" and "bound" β-amylases of barley. Characterization by two-dimensional immunoelectrophoresis, *J. Inst. Brew. London*, 84, 43, 1978.

106. Rowsell, E. V. and Good, L. J., Latent β-amylase in wheat, its mode of attachment to glutenin and its release, *Biochem. J.*, 84, 73, 1962.

107. Daussant, V. and Corrazier, P., Biosynthesis and modification of α- and β-amylases in germinating wheat seeds, *FEBS Lett.*, 7, 191, 1970.

108. Varner, J. E. and Ran Chandra, G., Hormonal control of enzyme synthesis in barley endosperm, *Proc. Natl. Acad. Sci. U.S.A.*, 52, 100, 1964.

109. Varner, J. E., Gibberellin controlled synthesis of α-amylase in barley endosperm, *Plant Physiol.*, 39, 418, 1964.

110. Filner, P. and Varner, J. E., A test for de novo synthesis of enzymes: density labeling with H_2O^{18} of barley α-amylase induced by gibberellic acid, *Proc. Natl. Acad. Sci. U.S.A.*, 58, 1520, 1967.

111. Jacobsen, J. V., Scandalios, J. G., and Varner, J. E., Multiple forms of amylase induced by gibberellic acid in isolated barley aleurone layers, *Plant Physiol.*, 45, 367, 1970.

112. Okamoto, K. and Akazawa, T., Enzymatic mechanism of starch breakdown in germinating rice seeds. VII. Amylase formation in the epithalium, *Plant Physiol.,* 63, 336, 1979.

113. Okamoto, K. and Akazawa, T., Enzymatic mechanism of starch breakdown in germinating rice seeds. IX. *De novo* synthesis of β-amylase, *Plant Physiol.,* 65, 81, 1980.

114. Swain, R. R. and Dekker, E. E., Seed germination studies. I. Purification and properties of an α-amylase from the cotyledons of germinating peas, *Biochim. Biophys. Acta,* 122, 75, 1966.

115. Swain, R. R. and Dekker, E. E., Pathways for starch degradation in germinating pea seedlings, *Biochim. Biophys. Acta,* 122, 87, 1966.

116. Abbot, I. R. and Matheson, N. K., Starch depletion in germinating wheat, wrinkled-seeded peas and senescing tobacco leaves, *Phytochemistry,* 11, 1261, 1972.

117. Julian, B. O. and Varner, J. E., Enzymic degradation of starch granules in the cotyledons of germinating peas, *Plant Physiol.,* 44, 886, 1965.

118. Shain, Y. and Mayer, A. M., Activation of enzymes during germination: Amylopectin 1,6-glucosidase in peas, *Physiol. Plant,* 21, 765, 1968.

119. Mayer, A. M. and Shain, Y., Zymogen granules in enzyme liberation and activation in pea seeds, *Science,* 162, 1283, 1968.

120. Cohen, E., Shain, Y., Ben-Shaul, Y., and Mayer, A. M., Structural and enzymatic characterization of plant zymogen bodies, *Can. J. Bot.,* 49, 2053, 1971.

121. Levi, C. and Gibbs, M., Starch degradation in chloroplasts, *Plant Physiol.,* 57, 933, 1976.

122. Peavey, D. G., Steup, M., and Gibbs, M., Characterization of starch breakdown in the intact spinach chloroplast, *Plant Physiol.,* 60, 305, 1977.

123. Levi, C. and Preiss, J., Amylopectin degradation in pea chloroplast extracts, *Plant Physiol.,* 61, 218, 1978.

124. Beck, E., Pongratz, P., and Reuter, I., The amylolytic system of isolated spinach chloroplasts and its role in the breakdown of assimilatory starch, in *Photosynthesis IV,* Akoyunoglou, G., Ed., Balaban Int. Sci. Serv., Philadelphia, 1981, 529.

125. Garwood, D. L. and Creech, R. G., Kernel phenotypes of *Zea mays* L. genotypes possessing one to four mutant genes, *Crop Sci.,* 12, 119, 1972.

126. Cameron, J. W. and Teas, J. H., Carbohydrate relationships in developing and mature endosperms of *brittle* and related maize genotypes, *Am. J. Bot.,* 41, 50, 1954.

127. Tsai, C. Y. and Nelson, O. E., Mutations at the *shrunken-4* locus in maize that produce three altered phosphorylases, *Genetics,* 61, 813, 1969.

128. Tsai, C. Y. and Nelson, O. E., Starch deficient maize mutant lacking adenosine diphosphate glucose pyrophosphorylase activity, *Science,* 151, 341, 1966.

129. Dickinson, D. B. and Preiss, J., Presence of ADP-glucose pyrophosphorylase in *shrunken-2* and *brittle-2* mutants of maize endosperm, *Plant Physiol.,* 44, 1058, 1967.

130. Weaver, S. H., Glover, D. V., and Tsai, C. Y., Nucleoside diphosphate glucose pyrophosphorylase isoenzymes of developing *normal, brittle-2* and *shrunken-2* endosperms of *Zea mays* L., *Crop Sci.,* 12, 510, 1972.

131. Hannah, L. C. and Nelson, O. E., Characterization of ADP-glucose pyrophosphorylase from *shrunken-2* and *brittle-2* mutants of maize, *Biochem. Genet.,* 14, 457, 1976.

132. Chourey, P. S. and Nelson, O. E., The enzymatic deficiency conditioned by the *shrunken-1* mutations in maize, *Biochem. Genet.,* 14, 1041, 1976.

133. Chourey, P. S., Starch mutants and their protein products, in *Maize for Biological Research,* Sheridan, W. F., Ed., Plant Molecular Biology Association, Charlottesville, Va., 1982, 129.

134. Nelson, O. E., Chourey, P. S., and Chang, M. T., Nucleoside diphosphate sugar-starch glucosyl transferase of *wx* starch granules, *Plant Physiol.,* 62, 383, 1978.

135. Burr, B. and Nelson, O. E., The phosphorylases of developing maize seeds, *Ann. N.Y. Acad. Sci.,* 210, 129, 1973.

136. Jarvi, A. J. and Eslick, R. F., Shrunken endosperm mutants in barley, *Crop Sci.,* 15, 363, 1975.

137. Kooistra, E., On the differences between growth and three types of wrinkled peas, *Euphytica,* 11, 357, 1962.

138. Collins, G. N., A new type of indian corn from China, *U.S. Bur. Plant Inds.,* 161, 31, 1909.

139. Parnell, F. R., Note on the detection of segregation by examination of the pollen of rice, *J. Genet.,* 11, 209, 1921.

140. Smith, L., Cytology and genetics of barley, *Bot. Rev.,* 17, 1, 1951.

141. Kempton, J. H., Waxy endosperms of corn and sorghum, *J. Hered.,* 12, 396, 1921.

142. Hixon, R. M. and Brimhall, B., Waxy cereals and red iodine starches, in *Starch and its Derivatives,* Radley, J. A., Ed., Chapman and Hall, London, 1968, 247.

143. Tomita, Y., Sugimoto, Y., Sakamoto, S., and Fuwa, H., Some properties of starches of grain amaranths and several millets, *J. Nutr. Sci. Vitaminol.,* 27, 471, 1981.

144. Wolf, M. J., MacMasters, M. M., and Rist, C. B., Some characteristics of the starches of three south american seeds used for foods, *Cereal Chem.,* 27, 219, 1950.

145. Irving, D. W., Betscharf, A. A., and Saunders, R. M., Morphological studies on *Amaranthus cruentes, J. Food Sci.,* 46, 1170, 1981.
146. Sugimoto, Y., Yamada, K., Sakamoto, S., and Fuwa, H., Some properties of normal- and waxy-type starches *Amaranthus hypochondriacus L., Starke,* 33, 112, 1981.
147. Okuno, K. and Sakaguchi, S., Inheritance of starch characteristics in perisperm of *Amaranthus hypochondriacus, J. Hered.,* 73, 467, 1982.
148. Murata, Y., Sagiyama, T., and Akazawa, T., Enzymic mechanism at starch in glutinous rice grains, *Biochem. Biophys. Res. Commun.,* 18, 371, 1965.
149. Akatsuka, T. and Nelson, O. E., Starch granule-bound adenosine diphosphate glucose-starch glucosyl transferases of maize seeds, *J. Biol. Chem.,* 241, 2280, 1966.
150. Akatsuka, T. and Nelson, O. E., Studies on starch synthesis in maize mutants, *J. Jpn. Soc. Starch Sci.,* 17, 99, 1969.
151. Tsai, C. Y., The function of the *waxy* locus in starch synthesis in maize endosperm, *Biochem. Genet.,* 11, 83, 1974.
152. Okuno, K., Gene dosage effect of waxy alleles on amylose content in endosperm starch in rice, *Jpn. J. Genet.,* 53, 219, 1978.
153. Echt, C. S. and Schwartz, D., Evidence for the inclusion of controlling elements within the structural gene at the *waxy* locus in maize, *Genetics,* 90, 275, 1981.
154. Vineyard, M. L. and Bear, R. P., Amylose content, *Maize Genet. Coop. Newsl.,* 26, 5, 1952.
155. Hilbert, G. E. and McMasters, M. M., Pea starch, a starch of high amylose content, *J. Biol. Chem.,* 162, 229, 1946.
156. Walker, J. T. and Merritt, N. R., Genetic control of abnormal starch granules and high amylose content in a mutant of Glacier barley, *Nature (London),* 221, 482, 1969.
157. Holder, D. G., Glover, D. V., and Shannon, J. C., Interaction of *shrunken-2* with five other carbohydrate genes in corn endosperm, *Crop Sci.,* 14, 643, 1974.
158. Preiss, J. and Boyer, C. D., Evidence for independent genetic control of the multiple forms of maize endosperm branching enzymes and starch synthases, in *Mechanisms of Polysaccharides Polymerization and Depolymerization,* Marshall, J. J., Ed., Academic Press, New York, 1980, 161.
159. Boyer, C. D. and Preiss, J., Evidence for independent genetic control of the multiple forms of maize endosperm branching enzymes and starch synthases, *Plant Physiol.,* 67, 1141, 1981.
160. Montgomery, E. M., Sexson, K. R., Dimler, R. J., and Senti, F. R., Physical properties and chemical structure of high amylose corn starch fractions, *Starke,* 16, 343, 1964.
161. Mercier, C., The fine structure of corn starches of varying amylose percentage: waxy, normal and amylomaize, *Starke,* 25, 78, 1973.
162. Taki, M., Hisamatsu, M., and Yamada, T., A novel type of corn starch from a strain of maize, *Starke,* 28, 153, 1976.
163. Boyer, C. D., Garwood, D. L., and Shannon, J. C., The interaction of the *amylose-extender* and *waxy* mutants of maize (*Zea mays* L.): fine structure of the *amylose-extender waxy* starch, *Starke,* 28, 405, 1976.
164. Boyer, C. D., Damewood, P. A., and Matters, G. L., Effect of gene dosage at high amylose loci on the properties of the amylopectin fractions of the starches, *Starke,* 32, 217, 1980.
165. Boyer, C. D. and Preiss, J., Multiple forms of starch branching enzyme of maize: evidence for independent genetic control, *Biochem. Biophys. Res. Commun.,* 80, 169, 1978.
166. Boyer, C. D., Garwood, D. L., and Shannon, J. C., Interaction of the *amylose-extender* and *waxy* mutants of maize: dosage effects, *J. Hered.,* 67, 209, 1976.
167. Fergason, V. L., Helm, J. L., and Zuber, M. S., Gene dosage effects at the *ae* locus on the amylose content in corn endosperm, *J. Hered.,* 57, 90, 1966.
168. Hedman, K. D. and Boyer, C. D., Gene dosage at the *amylose-extender* locus of maize: effects on the levels of starch branching enzymes, *Biochem. Genet.,* 20, 483, 1982.
169. Matheson, N. K., The (1-4)(1-6) glucans from sweet and normal corns, *Phytochemistry,* 14, 2017, 1975.
170. Boyer, C. D., Damewood, P. A., and Simpson, E. K. G., The possible relationship of starch and phytoglycogen in sweet corn. I. Characterization of particulate and soluble carbohydrates, *Starke,* 33, 125, 1981.
171. Karper, R. E. and Quimby, J. R., Sugary endosperm in sorghum, *J. Hered.,* 51, 121, 1967.
172. Singh, R., Effect of *high lysine (hl)* and *sugary (su)* mutant genes on improved nutritional quality of sorghum grain, Ph.D. thesis, Purdue University, West Lafayette, Ind., 1973.
173. Lavintman, N., The formation of branched glucans in sweet corn, *Arch. Biochem. Biophys.,* 116, 1, 1966.
174. Hodges, H. F., Creech, R. G., and Loerch, J. D., Biosynthesis of phytoglycogen in maize endosperm. The branching enzyme, *Biochim. Biophys. Acta,* 185, 70, 1969.

175. Black, R. C., Loerch, J. D., McArdle, F. J., and Creech, R. G., Genetic interactions affecting maize phytoglycogen and phytoglycogen forming branching enzyme, *Genetics,* 53, 661, 1966.
176. Boyer, C. D., Simpson, E. K. G., and Damewood, P. A., The possible relationship of starch and phytoglycogen in sweet corn. II. The role of branching enzyme I, *Starke,* 34, 81, 1982.
177. Bryce, W. E. and Nelson, O. E., Starch synthesizing enzymes in the endosperm and pollen of maize, *Plant Physiol.,* 63, 312, 1979.
178. Creech, R. G. and Kramer, H. H., A second region in maize for genetic fine structure studies, *Am. Nat.,* 95, 326, 1961.
179. Banks, W., Gerenwood, C. T., and Muir, D. D., Pollen starch from amylomaize, *Starke,* 23, 380, 1971.
180. Galinat, W. C., The evolution of sweet corn, Bull. 591, University of Massachusetts, Agric. Exp. Sta., 1971.
181. Yeh, J. Y., Garwood, D. L., and Shannon, J. C., Characterization of starch from endosperm mutants, *Starke,* 33, 222, 1981.
182. Boyer, C. D., Daniels, R. R., and Shannon, J. C., Starch granule (amyloplast) development in endosperm of several *Zea mays* L. genotypes affecting kernel polysaccharides, *Am. J. Bot.,* 64, 50, 1957.
183. Saussey, L. A., Morphological Changes in *Zea mays* L. Endosperm Conditioned by Mutants Altering Carbohydrate Composition, M.S. thesis, The Pennsylvania State University, University Park, Pa., 1978.
184. Fisher, M. B. and Boyer, C. D., unpublished.

Index

INDEX

A

Q

Q-enzyme, 138
Quantum efficiency, photosynthesis, 43, 48
Quantum requirements, C_3, C_4, and CAM plants, 80
Quantum yield, 45, 118—119

R

Rainforest plants, 45
Reaction centers
  light harvesting complexes, 38
  photosynthetic efficiency and, 39
  shade adaptation and, 46
Recombinant DNA, transformation, 30
Red algae, 38, 134
Red drop effect, 42
Reductive pentose phosphate cycle, 60, 75, 76
Regulators, photorespiratory, 110
R-enzymes, 140, 141
Reserve tissue, starch metabolism, 139—141
Rhodospirillum rubrum, 63, 64
Rhythms, starch metabolism, 134, 139
Ribulose-1,5-bisphosphate
  environmental regulation, 108
  photorespiratory pathway, 91
Ribulose-1,5-bisphosphate carboxylase/oxygenase
    (RubisCO), 48, 120
  activation, 65—66
  assimilate partitioning and, 6
  biochemistry, 61—65
    assay methods, 61—62
    catalytic properties, 63—65
    isolation methods, 61
    physical properties, 62—63
  biosynthesis, 66—67
  C_3-C_4 hybrids, 116—117
  C_3-C_4 intermediate species, 117
  C_3 plants, 75, 77
  C_4 plants, 78
  carbon assimilation, advantageous characteristics, 83
  enzymology, 112—113
  genetic engineering and breeding opportunities, 67—68
  inhibitors, oxygen, 97
  oxygenase-carboxylase ratio, variation in, 113—114
  photoinhibition and, 119
  photorespiration
    environmental factors and, 108—109
    flux rates and, 107
    photorespiratory pathway, 91
  photosynthetic rates and, 49—50
  practical considerations, 111—112
  somatic hybrids, tomato-potato, 21
Rice
  breeding, 5—7

carbon assimilation cycle, 75
starch metabolism
  degradation, 141
  mutants, 143
  in storage tissue, 139, 140
RNA, RubisCO, 112
Root crops, breeding, 7—8
Root development, in vitro, 19, 31
RubisCO, see Ribulose-1,5-bisphosphate carboxylase/oxygenase
Rye
  hybrids, embryo culture of, 29
  RubisCO kinetics, ploidy and, 64
Ryegrass, 113

S

Salpiglossis sinuata, 21
Seasonal photosynthesis, 51, 85
Sedum, 75
Seed, starch metabolism
  germination, degradation during, 141
  plastid integrity, 137
Seed development, sexual incompatibility and, 16
Seed maturation, O_2 depletion and, 93
Seed storage protein, nutritional quality and, 11
Seed yield
  canopy-apparent photosynthesis and, 51
  photosynthesis and, 85, 86
Self-pollination, 26
Semi-dwarf genes, 5—6
Serine
  as amino donor, 100
  Arabidopsis photorespiratory mutant, 107
  mutants affecting, 109, 110
  photorespiration biochemistry, 101—105
  photorespiratory pathways, 91
  variant selection, 116
Serine:glyoxylate aminotransferase
  amino donors, 100
  mutants, 104, 109
  photorespiratory pathway, 91
Serine hydroxymethylase, mutants affecting, 110
Serine transhydroxymethylase, 101—103, 111
  chloroplast isoenzyme, 105
  mutants, 107, 109
  photorespiratory pathway, 91
Setaria italica, 143
Sexual incompatibility, 16, 29
Shade
  crop yield and, 50, 51
  light reaction limitations and, 45—46
Shoot formation, in vitro, 19, 31
Shoot tip cultures, root and tuber crops, 8
Sink structures
  assimilate partitioning and, 5
  starch metabolism and, 145
Soil, root and tuber crops, 7—8
Solanum, 21, 22
Solanum tuberosum, see Potato
Solar energy, 9